Marriage, Gender, and Sex in a Contemporary Chinese Village

Marriage, Gender, and Sex in a Contemporary Chinese Village

Sun-pong Yuen, Pui-lam Law, and Yuk-ying Ho

Translated by Fong-ying Yu

An East Gate Book

M.E.Sharpe
Armonk, New York
London, England

An East Gate Book

The source of the main text of this translation is *Hunyin, Xingbie Yu Xing: Yi Ge Dang Dai Zhong Guo Nong Cun De Kao Cha.*
Copyright © 1998 by Global Publishing Co., Inc.

English translation rights arranged with Global Publishing Co., Inc., USA through Yuen Sun-pong.

English language edition copyright © 2004 by M.E. Sharpe, Inc.

Library of Congress Cataloging-in-Publication Data

Ruan, Xinbang.
    [Hun yin, xing bie yu xing. English.]
    Marriage, gender, and sex in a contemporary Chinese village / by Sun-pong Yuen, Pui-lam Law, and Yuk-ying Ho.
        p.cm.
    "An East gate book."
    Includes bibliographical references and index.
    ISBN 0-7656-1253-4 (alk. paper); ISBN 0-7656-1254-2 (pbk.: alk. paper)
    1. Women in rural development—China. 2. Marriage—China. I. Ruan, Xinbang. II. Luo, Peilin. III. Ho, Yuk-ying, 1964– IV. Title.
HQ1240.5.C6 R8313 2003
306.7′0951—dc21

                                                    2003011428

Printed in the United States of America

BM (c) 10 9 8 7 6 5 4 3 2 1
BM (p) 10 9 8 7 6 5 4 3 2 1

Contents

III. Sex and the Sex Trade Under the Reform and
Opening-Up Policy

* * *

IV. Epilogue: Baixiu Village at the Turn of the Century

Preface to the Expanded English Edition

From 1993 to the beginning of 2002, in a period spanning eight years, we conducted participant observation research in some villages along the Pearl River Delta. During the mid-1990s, we collated the research data and in 1998 published the findings in a book titled *Hunyin, Xingbie yu Xing Yige Dangdai Zhongguo Nongcun de Kaocha* (Marriage, Gender, and Sex in a Contemporary Chinese Village). The book explores the changing concepts on marriage and gender relationship, and attitudes toward sex among the Chinese of different cohorts over the past fifty years. It attempts to reveal in a wider perspective the tension between tradition and modernity against the economic modernization in the southern rural society in China with particular reference to the transformation of the Chinese self. Since the book was published, we have gone back periodically to the main fieldwork site of Baixiu Village and continued the research work. This expanded English edition is based on the Chinese edition, augmented by chapters 15 and 16, which are about the development of Baixiu Village at the turn of the millennium. Chapter 15 follows the path taken by the village and charts the social changes there over the last few years. Chapter 16 portrays mainly the political reforms taking place at the grassroots level, which the Chinese Communist Party put into effect. It deals with the election of the various village representatives. The emphasis is on how the grassroots political reforms impact the village society and the relationships among the villagers.

More than twenty years have passed since China's economic reform and opening-up policy was initiated in 1979. China has undergone changes that can only be characterized as cataclysmic. Market economic forces have swept China like an irresistible flood. While the Chinese Communist Party had sought without success to use political means to eliminate or alter traditional Chinese ways of life, the market economy has swept many such traditions away. This is clearly demonstrated in the development of Baixiu Village in the last two decades. Now, at the dawn of the twenty-first century, Baixiu Village can boast of visible developments in infrastructure, but economically and socially, advances no longer yield pleasant surprises to the observer. The village now exhibits most of the features of modern society and capitalism. The villagers' perceptions of life are basically no different from those of Chinese communities in capitalist societies like Hong Kong and Taiwan. Other than the appellation "socialist country," there are hardly any signs of a socialist or communist regime. On the other hand, many people have commented on one phenomenon in China's current development, namely, that reforms in the political sphere lag far behind those in the economic. Some even say that China's economic reform fixes its sights on only one thing—monetary profit—and that growth of the sense of individual autonomy of the Chinese is restricted to the sphere of economic self-interest. However, we are of the opinion that the question is more complicated than it appears. In chapter 16, the last chapter of this book, we shall attempt to explain the development of a kind of Chinese individual autonomy based on our examination of the grassroots political reforms in village society.

In December 1998, Baixiu Village held its first-ever one-person-one-vote direct election since the founding of the People's Republic of China, to elect the village representatives of the village assembly, the chairmen of six village groups (each group equivalent to a production team of the people's commune in the old days), the leader, deputy leader, and three members of the Village Committee. In the course of the election, even though there were various problems, the villagers participated ardently, maybe because it was their first opportunity to do so. The problems were varied. The Baixiu Village Election Committee did not follow all the election regulations in electing the group leaders. Some candidates offered bribes to get votes. Still, over 1,500 villagers voted, making for a turnout rate of over 95 percent. The Village Election Committee's work falls into two categories. The first consists of measures that implement the policies of the state. The Village Committee must follow the stipulated rules to "ensure the implementation" of the major policies of the central government such as birth control, tax collection, levying agricultural tax paid in grain, environmental protection, and so on. The second type of work pertains to the many undertakings related to the village—

public affairs ranging from its economy, to politics, to culture. The committee is supposed to deal with all such affairs by "democratic decision making" through "democratic management," under the critical "democratic supervision" of the village assembly. The governments of towns have no say in the affairs that are within the self-governing confines of the village; all they can do is to provide guidance, support, and assistance. In other words, the elected Village Committee has real executive powers to deal with the public affairs of the village.

The fact that the Chinese government institutes Village Committees throughout all the villages in the country is seen by many as the beginning of democracy in rural societies. Some researchers think that since the spread of the election of Village Committees in the late eighties, villagers have been able to practice the one-person-one-vote election to select members of the committee, so they have gradually learned the concepts associated with democracy. As a result, the basis for democratic development has gradually broadened.

Other researchers point out that even though the Village Committee has provided a chance for rural society to head toward democratization, the villagers themselves have not mastered the concepts of individual rights and individual autonomy. Thus it would not be easy for them to acquire the concept of political citizenship. Take the case of Baixiu Village. On appearance, the villagers went through the first election of the chairman and other members of the Village Committee, using the one-person-one-vote method, and the participation rate was very high. In actual fact though, to a large extent villagers' voting was influenced by factors such as their lineage. This means that in political reforms, peasants have not been able to demonstrate the individual autonomy of Western democratic societies. No doubt, some scholars attribute this to the present political system of China, which has not undergone any fundamental reform. The Communist Party is still the only ruling party. We should say that our main interest is not Chinese politics or the reform of its political system; it is, rather, through the election of the Village Committee, to assess the development of the self-concept of the Chinese, in particular the inhabitants of Baixiu Village. As pointed out by some scholars, traditional Chinese culture poses an obstacle to the growth of the villagers' sense of individual rights and individual autonomy and similar ideas. In the rest of the chapter, we shall approach this claim from three angles: Chinese familial culture, an individual's economic self-interest, and autonomy. Put another way, we shall chart the development of self-concept in contemporary China by way of this ground-breaking election, and shall focus on the concept of autonomy.

As mentioned in the preface to the Chinese edition, chapters 1 through 14 were written by different authors, yet they were discussed thoroughly by all three authors. In this English edition, the newly added chapters 15 and 16 were initially drafted by Yuk-ying Ho and Pui-lam Law, respectively, while the final drafts were the product of the collective thoughts of the three authors.

February 2003

Preface to
the Chinese Edition

This book is a report of the findings of a social research study of a village in the Pearl River Delta in Guangdong Province, China. Its purpose is to reveal the changes in views about marriage, the relationship between the sexes, and attitudes toward sex in a village in southern China over a period of about half a century. The study's main thrust is to attempt to integrate the concepts of Chinese familism, Western feminism, and the development of individual autonomy into a theoretical framework. The investigation employed the research methods of participant observation and thick description to gather, narrate, and analyze data. At the same time, we approach the analysis from the angle of critical hermeneutics, setting it against the evolved theoretical framework, so as to view and evaluate the development of events and related social phenomena.

Participant observation is a method frequently used in social science, especially anthropology, but in the study of the Chinese village, in particular studies written in Chinese, the use of thick description to present data is comparatively rare.[1] One of the features of this book is to use stories and intensive interviews to depict some of the changes taking place in the contemporary Chinese village. Our discussion will show how the Chinese self-concept changes during the process of modernization and adapts to it.

Using change in self-concept as a point of entry into our research will highlight two issues: the indigenization of social research and the perspective of critical hermeneutics. The former refers to the danger of narrowly applying Western theories to the study of Chinese society. The risk is that such theories are premised on the Western conception of the individual and presuppositions about the relationship between the individual and society. Neither premise sits well in the fabric of Chinese society. The latter issue, that of critical hermeneutics, precisely because it embodies Western conception of the individual and presuppositions about the relationship between the individual and society, throws a light on the contrasts, and even casts a critical look at the "reasonableness" of social practice in present-day China, especially in the way traditional culture and the social system still constrain the individual.

To understand self-concept and its dilemma in its indigenous context through critical hermeneutics has been one of the main concerns of our recent research in the past few years into the cultural and social changes in China. This book is a product of that endeavor. We should add, however, that this is only an exploration, whether in terms of empirical investigation or theory construction. Our research is proceeding at different levels, and we hope to continue to publish the results of both theoretical elaborations and empirical studies.

Data for this research report were mainly gathered in 1994 and 1995. Five workers were involved. We rented a flat in the village where we conducted our fieldwork, and used the method of participant observation to conduct our study. As we carried on our regular teaching commitments while collecting data, we spent a few days at a time during weekends and on nonteaching days to conduct the research and then rushed back to Hong Kong to resume teaching. The collection of data was carried out intermittently; this added to the problem of adaptation to changing environments. We found that at times it was tricky to alternate between Hong Kong and the field, since the social environment and the associated human relations changed so much. We are therefore all the more thankful to the assistance of our informants in Baixiu (White Grace) Village, where our fieldwork took place.[2] They not only provided us with information, they also not infrequently looked after our daily needs and our personal safety in a deteriorating environment. We want to express our special thanks to our department head, of the Department of Applied Social Sciences, Professor Diana Mak. It is she who has given this research encouragement and support, especially in the arrangement of teaching duties and facilitation of the study. Finally, we thank the Hong Kong Polytechnic University for grants in support of this study.

The data collection work was shared in the following manner: Other than the data about karaoke sing-along girls in nightclubs, which were collected by the male researchers, the other parts were divided so as to match the gender of the research group members with the gender of the interviewees.

The data on women workers were collected by our research assistant, Li Jing. In the narration, the pronoun "I" is the interviewer. Although the chapters were written by different authors, they were discussed again and again by all three. Sun-pong Yuen was responsible for chapter 1 and chapters 10 through 14; Pui-lam Law for chapters 2 through 6; and Yuk-ying Ho for chapters 7 through 9.

February 1998

Marriage, Gender, and Sex in a Contemporary Chinese Village

1

Theoretical Framework

Familism, Women's Liberation, and Autonomy

The aim of this book is to investigate the changes in views about marriage, relationship between the sexes, and attitudes toward sex in the rural villages in southern China. The research mainly adopted the method of participant observation to conduct an empirical study of a village in the Pearl River Delta. The data will be presented in the form of narrative stories. We hope that by way of thick description, we can present those facets of changes that have occurred in Mainland China in the last half-century, in particular the last twenty years or so. Through this discussion, we hope to reveal some of the progress and the setbacks that China encounters in the march toward modernization.

The book is divided into three parts. Part I consists of the stories of three marriages at three different points in time and investigates the problems of finding a spouse, conducting a courtship, and maintaining a marriage through the eyes of various men. The three weddings took place, respectively, before 1949, in the late 1970s, and at the beginning of the 1990s. The cases represent changes in views about marriage and related matters in the villages in southern China, through a process of transition from tradition to modernity.

Part II investigates how women of different generations perceive self-actualization during the transition period from tradition to modernity. The protagonist of the story in this part is a young woman who grew up in a village during the time of the reform and opening-up policy after 1979. An old woman and a middle-aged woman will also figure as contrasts to the protagonist in her quest for self-understanding.

The theme of Part III is the relationship between the sexes. The foci are the sexual relationship between men and women and the sex trade in the village. This part delineates the economic prosperity in the Pearl River Delta brought about by foreign investments and the consequent "abnormal" sexual relationships and sex trade.

At a more general level, this book deals with changes in ideas about marriage and relationships between the sexes in the context of "modernization." We have employed some indigenous Chinese concepts and some Western concepts in our interpretive and analytical framework. First, we take up the concept of Chinese familism, with particular reference to its characteristics such as the male-dominated "patriarchal culture," the male–female relationship expressed by the dictum "men superior, women inferior," and the value of "ultimate concern." Second, we make use of some of the ideas critical of the relationship between the sexes espoused by the women's liberation movement in the West, a movement that was initiated in the twenties and reappeared as a second wave in the sixties. Last, we employ a group of theories that relate the development of individual autonomy and consciousness of selfhood to modernization driven by commodities or market economy. Furthermore, we take a critical stance in understanding how market-driven autonomy weakens the value of "ultimate concern" and examine how this may generate the dissolution of traditional social order in contemporary China. In using the above theories as a framework to interpret our data, we are careful not to impose Western theories on Chinese situations. We attempt to combine the application of Western theories with the context of Chinese traditional culture and historical development in our research.

Two reasons motivated us to use the above-mentioned Western theories as our interpretive and analytical framework. In the first place, this research is concerned with changes in views about marriage and the relationship between the sexes within the context of modernization. Western social sciences have had a rather rich tradition of theorizing about the transition from tradition to modernity. Any study that explores the relationship between tradition and modernity must perforce refer to Western social science theories. Second, although we realize that the development of Western social science theories took place within the corresponding context of the histories and realities of the West, and it is not possible to apply them to Chinese situations in their entirety, nevertheless the historical phenomena and social contexts that the Western social theories attempt to explain serve some useful purposes in relation to the direction of Chinese modernization and the present-day Chinese social system. They allow us to see more acutely, to contrast, and to some extent to evaluate the "legitimacy" of the latter. In this sense, the deployment of Western theories as an analytical tool also serves a "critical" function. Overall, then, we attempt to combine Chinese theories with relevant Western theories and to adopt both empathetic and critical perspectives so as to explore some of the developments and setbacks that China experienced along the road to modernization.

One point needs to be noted before we launch into an exposition of our theoretical framework. The main period under study is post-1979 China. Scholars of contemporary China will agree that this is a time when the old gives way to the new, a time of cataclysmic changes, a time when both traditional culture and modern values come to bear on a person's behavior. In short, it is a time when paradoxes and conflicts at both the individual and social levels abound. Three integrated sets of theories from the Western and Chinese academic traditions will be used for analysis at different conceptual levels and angles. Sometimes the same person's behaviors and events may be described and analyzed using different or even opposing theoretical concepts. This may create an impression of inconsistency; however, it is difficult to avoid such an impression. It aptly reflects, we think, the confusion experienced by China at a crossroads.

Familism

No one will deny that familism or familial culture lies at the core of traditional Chinese society. On the other hand, the changes in the social, political, and economic aspects of China in the past hundred years have been tremendous, and manifest themselves in changes in the Chinese way of life. China, Taiwan, and Hong Kong, whose populations are predominantly Chinese, show markedly diverse patterns in their political, social, and cultural facets. The changes have been most noticeable in Mainland China, where there has been a deliberate attempt to eradicate traditional values and ways by means of sundry political movements and economic reforms. As for Hong Kong and Taiwan, although exempt from the baptism of violent political changes, lifestyles there have nonetheless undergone considerable change, affected in many ways by Western values. One might ask the following: Is the impact of familial culture as strong as in the past? Is familism still a good starting point for the study of the situation of contemporary Chinese?

In the past twenty to thirty years, familism remained as the main research anchor in most studies of the Chinese way of life. It is undeniable that the style of living of the modern Chinese departs from that of the traditional, but familism and its value system still affect greatly the behavior of Chinese everywhere.[1] Our study centers on the Chinese village after the reforms and opening-up policy of China starting in 1979. In fact, the post-1979 period is generally regarded as one of reawakening for familism in China. With the implicit approval of the central government, pre-1949 lineage activities revived. Ironically, under the impact of the market economy, the rise of individual autonomy counters the values of traditional familism in many ways. For this reason, some studies have pointed out that it is precisely during this period that the effect of familism on the individual weakened day by day.

This appears to be a contradiction. But if we observe carefully, we will see that far from being eradicated, familism has continued to influence people's behavior in the past decades. Only the related rituals and activities were suppressed. When the activities and rituals were revived, they could have exerted a stronger influence on people's behavior. However, as we said earlier, some of the values of the market economy were to a large extent at odds with those of familism; thus the influence of familism has been correspondingly weakened, especially in the southern provinces. This is an illustration of the tension between tradition and modernity.

We can approach an understanding of familism at two levels: at the more concrete or human relationship level and at the more abstract or general level. The more concrete level denotes some criteria of behavioral standards. We can use the "five human relationships"—between sovereign and subject, father and son, husband and wife, among brothers, and among friends—as the focus of analysis. These five relationships form the basis of proper behavior for social and individual behavior. Of course, in the modern world, the field of sovereign and subject no longer applies, or needs to be invested with new meaning. What is significant here, though, is the kind of "patriarchal culture" and its related human relationships. We will make use of Francis Hsu's concept of the "father–son dyad" and Fei Hsiaotung's concept of "differential mode of association" as explanatory tools.[2] Familism or lineage culture puts the "father–son dyad" at the center of all other relationships, and extends the blood relationship of paternity outward to other relationships and so constitutes an extension of familism. Within the structure of a lineage, familism—the value system that prescribes behavioral standards—requires that members of the lineage act in accordance with the ultimate goal of continuing and expanding the lineage. On the level of decision making, what guides decisions under familism revolves around the axis of the father–son relationship. The latter highlights one of the important characteristics of familism: patriarchal culture, and out of that the male–female relationship of men dominant, women subservient.

From the point of view of the structure of a family or a lineage, familism entails two interrelated human relationships. One is the father–son relationship with its strict differentiated order according to seniority; the other is the relationship obtaining between the sexes, which holds that men are superior and women inferior. Viewed from these two angles, although men, who are already in a leading position, should respect women who are more senior than they are; in actual fact, when it comes to decision making and authority, men's status is higher than that of women of a senior generation. Starting from the "father–son dyad," and extending outward according to the "differential mode of association," the Chinese human relationship network is formed.

The second, more abstract level of approaching familism is to look for some general characteristics or attributes exhibited by the specific behavioral standards espoused by familism. These attributes can be taken to be the deep structure of Chinese culture. We can liken them to the attributes of Francis Hsu's "father–son dyad": continuity, inclusiveness, authority, and asexuality.[3] According to Hsu, a relationship network revolving around the axis of the father–son relationship will naturally emphasize the quality of continuity. The father–son blood relationship is one that is stable and will not be changed by internal or external factors. If all other relationships, such as husband and wife, and friendship, are also subsumed under the quality of continuity, then the mark of inclusiveness in Chinese human relationships is apparent. Fei's construct of "differential mode of association" neatly explains the practice of defining relationships from the closest to the most distant with kinship terms that are applicable to relatives and brothers. This is a mark of inclusiveness. Although relationships are seen as closer or more distant, because they center on the father–son relationship, they exhibit to varying extents the attribute of authority whereby the seniors are superior and the juniors inferior. The same holds true for the bond between husband and wife or the relationship between men and women; the father–son dyad invests in them the attribute of asexuality, thus ignoring the intrinsic differences in gender between men and women.[4] At the macro level, the above analysis serves to reveal the deep structure of Chinese culture. At the micro level, it reveals the character or self-concept of the Chinese.

We could combine Francis Hsu's concept of "father–son dyad" and Fei's "differential mode of association" to better examine the behavior of the Chinese. There are two strata of human relationships in this integrated view. One is to start from the "father–son dyad," that is, to start from the family to the lineage, and go thence to friends and acquaintances, thus constituting a network of relationships in which the polar relationships of close versus distant and superior versus inferior are observed. In other words, every person's role as well as status is assessed from the point of view of the family, in terms of the ranking the person achieves in seniority and in gender. In the relationship between the sexes, the rule of "men superior, women inferior" applies.

At the second stratum, the defined relationships become the basis for the distribution of resources and benefits, especially in regard to wealth and jobs.

Equipped with the above analysis, we can compare the differences between the Chinese conception about human relationships and that of Westerners; the contrast will sharpen our understanding of the Chinese character. One main difference in the conception of human relationships between the Chinese and the Westerners is that the former emphasize the continuity of a relationship whereas the latter emphasize its function. From the Chinese point of view, the bond between father and son passes from generation to generation: it is everlasting. In assessing, interpreting, and understanding other relationships in terms of this pivotal bond, what is important is continuity. In contrast, for Westerners, the relationships of husband and wife, father and son beyond adulthood, and among family members become weakened, with respect to duties and responsibilities, past a certain stage in life. Westerners, especially citizens of the United States, throughout their lives, associate themselves with a number of societies and organizations. On the whole, such societies are functional and are joined to meet the various needs arising at the different stages of life.[5]

Close scrutiny will show that the value system underlying a continuity-based relationship is inclined toward conforming to group values and using the degree of closeness of relationship in the "differential mode of association" as the bases for value judgment and distribution of benefits. In contrast, the value system emphasizing functional orientation in relationships makes self-interest the compass for one's behavior and the principle of fairness the criterion for the distribution of resources. The two value systems will naturally give rise to two modes of behavior and different criteria for value judgment and orientation. This in turn gives rise to two different concepts of the self. To simplify, because the mode of behavior and value orientation of the Chinese are based on the well-being of the collectivity of the lineage, their self-concept is hardly distinguishable from the concept of the "larger self," that is, the larger groups formed by relatives and lineage members. Personal likes and dislikes are limited by the concept of the larger self. This is very different from the attitude of Westerners, who see self-interest as the guiding principle. True, their self-concept is also influenced by the societal larger self, but in deciding on the priority of actions and benefits, the individual self will take precedence over the larger self.

Applying the above analysis to the topic of marriage and associated matters, we can clearly see that the patriarchal culture and the "men decide, women follow" relationship between the sexes, which is characteristic of familism, regulate the behavior of the Chinese in their courtship and marriage customs. In traditional China, marriage was not based on affection, but rather on the interest of lineage. Viewing the traditional Chinese concept of marriage through the lens of modern-day values, we can see that in such matters, individuals are powerless to make their own decisions. We can even say that the behavioral standards and criteria according to which a traditional Chinese behaves are not contingent upon individual "likes and dislikes." The individual's desires are all subjugated to the "larger self" of the family and the lineage.[6] If one compares the lives of men and women, taking the father–son dyad as the principle into which to subsume all other human relationships, then it is clear that women face many more constraints than men in matters of marriage and daily life and behavior.

To deal with the people and the issues of our research, we employ the concept of father–son dyad and its extension, the relational network of "differential mode of association," as our analytical framework. The relationship between the sexes characterized by "men decide, women follow" and "men superior, women inferior" is the focus of our analysis. We also pay attention to the tension between familism and the development of individual autonomy. We further examine the idea of "ultimate concern" generated by Chinese familism. In brief, "ultimate concern" here refers to a value of familial concern transcending the individual and extending to kinsmen and friends, and emanating from which a social network based on family and lineage ties is woven. Based upon this understanding, the relationship between "ultimate concern" and the dissolution of social order will then be explored as well. These issues will be discussed in greater detail in sections two and three of this chapter and in chapter 13.

"Men decide, women follow" was the model for the relationship between the two sexes in traditional Chinese society. The relationship between the two sexes and the idea about marriage before 1949 belonged, of course, to the traditional model. After the Communist Party took over the ruling of China, a new marriage law was promulgated in 1950. A series of laws concerning women were enacted with a view to obliterating the inequality between the sexes. That these measures actually led to the equality of the sexes is very much doubted by those who have studied the question. After the implementation of the reform and opening-up policy, many traditional social practices and lineage activities showed signs of revival, especially in Guangdong's Pearl River Delta. What are the main differences between the revived lineage activities and those before 1949? This is a most interesting question, one that awaits closer scrutiny. From our field study, one thing is quite clear: the traditional model of the relationship between the sexes has resurfaced, and in some respects has shown signs of becoming more entrenched. One main factor contributing to this is the economic environment and not lineage activities per se. With the development of a market economy, China has experienced an economic boom, especially in the area of the Pearl River Delta adjacent to Hong Kong. Investment flowed in from Hong Kong, Taiwan, and foreign investors. Many factories were built, and female workers from outside the area poured in to fill low-level jobs in these factories. The functional structure of labor in such places replicates the "men decide, women follow" theme. Moreover, sex trade thrived in the area. The consequence is that the dormant "men decide, women follow" relationship reared its head once again.

The stories of the three marriages in different generations and the description of the sexual phenomena in Parts I and III of this book, respectively, show this male-dominated relationship clearly, and indirectly provide strong confirmation of the characterization of the "father–son dyad" of lineage culture. However, looking at the development of the market economy and commercial activities, we detect the emergence of something that appears to run counter to familism, and that is the development of individual autonomy. The three marriages described in the book reveal the fact that the younger the generation, the more the couple respects equality for each other. The story of the young woman in Part II shows that as women pursue a career, they develop a comparatively more independent self-concept. There is in these phenomena evidence of tension between lineage culture and the development of individual autonomy.

We can rephrase the question about lineage culture like this: Does the revival of lineage culture after 1979 strengthen its influence on the individual or, faced with the market economy, is its influence weakening? There is no simple direct answer. From data gathered in this study, we might characterize the present state of affairs in this way: After 1979, two contradictory tendencies concerning lineage culture are discernable. Viewed historically, the revival of lineage activities means that the political movements and economic reforms of the past several decades have not eradicated them. Yet while such lineage activities appear to have received recognition once again, social problems have arisen that have become so serious that a sense of "dissolution of the social order" has set in. Some important concepts and constraints in traditional familism appear to have lost their effectiveness in directing the behavior of contemporary Chinese people. The feeling of "ultimate concern" that familism used to instill in people is thinning. This is all too obvious when we come to Part III of the book, about sex and sex trade in the area. Thus we arrive at one of the paradoxes in China's road to modernity. The power of familism has emerged from hiding. Paradoxically, some of its main features become weakened by the prosperity brought about by the market economy.

Western Women's Liberation Perspectives

One of the characteristics of the relationship between the sexes under traditional Chinese familism is "men decide, women follow," or "men superior,
women inferior." Such a tradition forms the backdrop against which we can
observe the relationship between the sexes and other developments in women
after 1949. The leadership and some scholars in China consider the post-
1949 period the one in which women became liberated. Therefore, in this
research, we employ some notions developed by the Western feminist movement to examine the condition of contemporary Chinese women and assess
the extent of their liberation. We are particularly concerned with post-1979
developments. Although the market economy has in some ways aggravated
the inequality between the sexes, it has also fostered the emergence of individual autonomy, which in turn sparks the awakening of the female consciousness. In this sense, appealing to the Western notions of women's rights
as an analytical framework serves a useful purpose.

Quite a number of Western scholars, especially those advocating
women's rights, have pointed out that the laws concerning the protection
of women passed after 1949 were passed not to advance the rights of women
themselves but for economic benefits. At different times and for economic
purposes, the country passed laws and glorified women's roles on the
premise that women could play a contributory role in the country's
economy. More important, various land reform movements and economic
planning—for example the distribution of land and the system of work
points—were based on the family as the unit of measure. Specifically, the
head of the household was the recipient of any economic benefits that
went to the family unit. In this way, the scholars argued, land reform and
collectivization had the opposite effect of consolidating the male-centered
patriarchal culture and the entailed relationship between the sexes.[7] Some
scholars argue further that subsequent to Deng Xiaoping's reform and opening-up policy in 1979, the rise of the market economy has worked to make
women lose ground in terms of equality with men. During the Mao Zedong
era, many political movements encouraged women and men to participate
side by side. This improved the chances of equality for men and women.
Now, in the period of open-door policy, even though women enjoy more
opportunities to engage in economic activities alongside men, they generally take low-level jobs in corporations and factories. There are few political and social movements for them to take part in, unlike the days of
Mao. The result is that women's dependency on men has become more
pronounced, and this retards the development of a more equal relationship between the sexes.[8]

These scholars' views are typical of Western feminist theories that interpret the situation of Chinese women today. Such a perspective undoubtedly
offers insights and new understanding. However, women's issues in China
have been developed in their unique historical context. They are more complex than they appear. In the following paragraphs, we shall first delineate
some key viewpoints of Western feminism and their strengths, and then use
them to examine the Chinese situation.

Feminism in the West developed in the 1920s and 1930s. At the end of the
1960s, when the anti–Vietnam War movement was at its height, another
women's rights movement was on the rise.[9] This time, not only did its influence spread all over the world, but the type of rights demanded were more
uncompromising. Now after twenty years of real-life struggle and theoretical exploration, Western scholars have gained a thorough understanding of
how Western women suffer discrimination. They are able to deploy multiple
perspectives in their assessment, such as the historical developmental context, constraints imposed by the social system, and the influence of ideology.
With these, they analyze women's situation, assess the relationship between
the sexes, and severely criticize men's domination over women in all spheres
of life. Such analyses are manifestations of Western scholars' deep knowledge of women's existential condition.

Quite a number of feminist activists and scholars of women's status in
society not only direct their criticisms at the tangible evidence of male domination, but extend their critique to ideological domination, the structure of
sociopolitical theories, and even modes of thinking and methodology. For
these thinkers, the inequality of the sexes is traceable not only to the fabric of
the historical process but also to the very theory construction and thought
process that are rooted in a male sensibility.[10] Typical of such criticism in
Western social and political thinking is the discussion of public space and
private domain. The former refers to public affairs; the approach to such
affairs should be selfless and fair. The latter refers mostly to affairs concerning family life; what it involves is sensibility and private domain. Operating
in these two different spheres are different modes of thought: the former
requires a mode of abstract logical thinking, while the latter involves piece-
meal intuitive thinking.[11]

The women's liberation movement in the early twentieth century targeted
mainly clear and easily identifiable instances of male oppression of women
in both the public and private domains. The second wave of the women's
liberation movement and women's study from the 1960s onward, on one
hand, continues the first wave and, on the other hand, goes much further in
launching a critique of male domination in society at the level of theory
construction and mode of thinking in social and political inquiry. Feminists

point out that this apparently gender-neutral and universal theory construction and research direction reflects what men's value system and attitude toward life have been historically. Whether out of nature or nurture, women's orientation toward research questions, their mode of thinking, and their construction of the self-concept are not in tune with mainstream social studies and are even suppressed in extreme cases.[12] If we study Chinese women from the perspective of the development of women's liberation, what important insights can we gain?

We can approach this by pursuing two lines of inquiry: one, the tangible course of historical events, and the other, its deeper structure. Regarding tangible historical development, many scholars have pointed out that, although women's position in China after 1949 has improved compared to the one that preceded, women have been far less liberated than the government has claimed and women's studies scholars of the Marxist school had hoped. As mentioned earlier, the policies on women during this period were not formulated with women's interests in mind, but with economic development and social stability in mind instead. The fact is that women living in this period had little inkling, in their consciousness and in daily affairs, that they should have the same rights as men. Nor were they aware, in any tangible way, that women were oppressed. The so-called women's awakening, or the higher status of women, was only promoted and publicized as a strategy to achieve economic or political ends. The pervasive influence of traditional Chinese culture and social structure on women, especially in the construction of their self-concept, has not been fundamentally altered despite political and economic changes.

This observation leads to study of the issue at a deeper level. The relationship between the sexes—men superior, women inferior—is a product of 2,000 years of familism. What people in Mainland China can spot are mainly some concrete instances of inequality between men and women. In the past decades in China, local scholars who studied women as well as advocates of women's rights could only analyze such surface phenomena. They did not probe either the level of theory construction and mode of thinking in evaluating women's condition or the root of women's oppression. The fact is, to a large extent, women's studies scholars in China have not come to grips with how the construction of social and political theories and the mode of thinking have led to the position of male dominance.[13]

What we ought to be careful about is that such theories and analyses stem from a Western cultural context and presuppose Western cultural views about "individual," "self-concept," and "individual's relation to society." In other words, when we apply the theories of feminism to China, we are also using their implicit conceptions of the individual in the analysis of Chinese women.

This further implies that value-laden terms such as "self-determination" and "self-awareness" are used to understand and even evaluate the existential situation of Chinese women. Indeed viewing Chinese women from such value perspectives has the advantages of contrast and distance. We can also, to some extent, assess the reasonableness of the social system in which Chinese women live. Nevertheless, we are also fully aware that the analysis and assessment of China's conditions should be made within their own social context.

The population of China throughout its long history consisted of over 80 percent peasants. They lived in a stable social structure born out of familism. For two thousand to three thousand years, neither the chance nor the conditions existed for men or women to develop an independent individualistic character, let alone for a woman's self-consciousness to be awakened. After 1949, new economic and political policies had a huge impact on the peasants' traditional way of life. And new laws concerning women were passed. Comparatively speaking, women were better off than before 1949. After the reform and opening-up policy of 1979, a market economy developed. Contrary to what some Western scholars have argued, this development has not pushed women back to the family circle, or blocked the development of their autonomy. If we look at Chinese women from the perspective of the slow march of historical development, we see that, both in the country and in the city, women's rates of employment and the numbers employed have increased significantly. Because of the improvement of the economy, people's income has increased. Women have come to gain greater control over earned resources, and this has indirectly led to the development of their individual autonomy.

There is one point that needs to be emphasized. If we hold that the rights enjoyed by Western women are, to a lesser or greater extent, the goals that Chinese women are striving to achieve, then to attain Western women's awakened consciousness and degree of self-understanding will require the conjoining of the right economic, political, and social conditions. In post-1979 China, the chances of and conditions for women's development are far greater than before. However, this is not to deny the insights into the conditions of Chinese women after 1979 advanced by Western scholars. To the contrary, our observation is that in the mid-1990s, the market economy, the structure of labor in corporations and industries, and especially in the prosperous sex trade in the Pearl River Delta all show that women are oppressed by a new form of inequality. From a broad perspective, we can see the paradox of China's modernization led by a market economy. The provision of material needs has improved, and the vista of everyday life has widened. At the same time, the developments have engendered even more oppression and inequality than found in capitalist countries. This paradox is very much in evidence in the development of the relationship between the sexes. The Chinese woman in the 1990s enjoyed conditions much more favorable to fostering self-determination, as well as for achieving equality of the sexes. Her economic conditions, her life experience, and her outlook were enriched. Yet, the emergence of patriarchal organization structure and the booming sex trade accompanying the market economy led to a serious degree of commodification of women.

These are two opposing trends in women's condition in China. In Part II of this book, we present the case of a young woman's self-understanding. And in Part III, we present the sex trade in Baixiu Village. These cases show two scenarios of a woman's life: one with opportunities for betterment and one with oppression. This is a paradox caused by the Chinese modernization process. We shall not resolve the paradox by any expatiation. We would point out that the key to the puzzle lies in how we view the status of Chinese women during the period of 1949 to 1979 compared with the period after 1979. Obviously, women's lives today have improved over those of traditional times, even with the commodification and the oppression; but how do we compare the pre- and post-1979 periods under communism? Here we have two opposing trends of development, and two different perspectives to analysis, that is, the Western feminist angle and the context of China's own historical development. If we can integrate Western feminist theories with the context of China's traditional and present realities, then we should be able to gain a more thorough and more accurate understanding of Chinese women's present condition and future prospects.

Autonomy, Ultimate Concern, and the Dissolution of Social Order

From the above discussion of theoretical issues, one will gather that this book adopts two opposite frameworks of analysis. In one, we take the characteristics of familism and assess the constraints they place on the individual, views about marriage, and gender relationships. In the other, from the vantage point of Western women's liberation and the development of individual autonomy, we examine the degree of women's initiative in matters concerning marriage and opportunities for self-fulfillment.

These two contrasting analytical frameworks are used in the stories in the first and second parts of the book. The story of a marriage before 1949 and a woman growing up in that period shows clearly how familism conferred a sense of stability on people's value system but created, to varying extents, misery in the lives of men and women.

Since 1979, China has undergone momentous changes in its social and economic spheres. The political movements that took place in China for over thirty years came to a halt. A market economy was formed; its impact on people's lives was increasing. In addition, people came to the painful realization of the fickleness of the ruling methods and policies of the governing class in the past. The ideology of the Communist Party has been on the decline among the people and even among party members. The market economy has vastly improved people's standard of living, especially in the southern provinces of China, and more particularly in the Pearl River Delta. Hong Kong, Taiwan and other foreign investors built large numbers of factories, igniting economic development and changing people's way of life both materially and socially. This gives the people in those regions greater self-determination and power over their acquired resources and thus a greater say in their own destinies.

In this book, we have examples of middle-aged men who have considerable freedom in choosing a wife. They also know how to make good use of the open-door opportunity to improve their standard of living. Yet traditional familial culture still imposes a standard code of behavior on men of this generation. For them, "men superior, women inferior" is still the guiding model for familial relationship.

For the young people of a new generation, particularly those with more education, autonomy in daily life is even more evident. In an example of a marriage in the 1990s described in this book, more respect is shown between husband and wife than in the older generations. Women now have a greater say in how they choose to use material resources. We will see how a woman quite independently decides the various matters in her daily life and how she plans and what she hopes for the future.

However, historical development does not proceed in a simple linear manner toward the good or the bad. China in the 1990s had a more prosperous economy, and people's standard of living improved a great deal. People could make decisions about the affairs in their daily lives according to their own will. Men and women alike had a much greater degree of choice and a far more expansive view of life compared with those in the past. Does all this mean that China is moving toward a proper course of modernization?

To many academics who study the history of modern China, the answer to this question is most certainly not straightforward. To look at political reform for instance, one can see quite clearly that political reform has not kept up with economic and social changes. Not only that, it contains seeds of sharp conflict and discord.[14] In Part III, we look at the thriving sex trade that came about in the late 1980s in the Pearl River Delta and the consequent aberrant relationship between the sexes. It is from this vantage point that we attempt to answer the question posed in the previous paragraph.

In the area where our fieldwork was conducted, we note that almost all married men who are sexually active have extramarital sex or have visited sex-related establishments. There are nightclubs that provide girls for company and barber shops offering "special services." The part of this book that deals with this topic will recount the stories told by people of different social strata from both sexes, and the stories tell of the seriousness of the problem.

Instrumental values, chasing of wealth, and profligacy pervade the entire Pearl River Delta. They might well be common to all modern societies driven by a market economy. Yet from our observation, in the Pearl River Delta, or more precisely the area where we conducted our fieldwork, chasing wealth and indulging in sexual pleasure seem to be the main goals of life for most men there. In a society so intent on pursuing sexual gratification, the mutual care that the family and other associative relationships foster has thinned out. But is the kind of "ultimate concern" promoted by familial culture entirely gone? To answer this question, we can bring out the third theoretical framework employed in this study: individual autonomy and the idea of "ultimate concern" closely associated with familism.

One of the hallmarks of a modern society driven by a market economy is the respect accorded personal choice and the priority given to autonomy. Taking a broad perspective, we can say that that is the goal Western society has strived to achieve and, to a large extent, has achieved. If we then look at post-1979 China from that perspective, and examine the effect of the market economy on the development of individual autonomy during that period, we will see the present situation in China in a clearer light. We will come to a better understanding of the relationship between the development of individual autonomy, market economy, and familism. But first we should ask, What is autonomy? Does it mean different things in Chinese and Western cultures?

Autonomy, in relation to Western individualism, is a state or capacity for the self-initiation of thoughts and actions. People who are autonomous can make decisions about their actions on their own, free from external constraint.[15] "External constraint" implies the use of force. For example, people under an authoritarian regime are afraid to express their own free will. We can say in such circumstances that they possess the capacity for autonomy but dare not express it.

A second way of looking at autonomy is to see it in relation to "internal constraint." Internal constraint can refer to social or moral norms that over a long period of time imperceptibly influence and then control one's behavior. We should point out, though, that this notion is contentious. Some people would argue that any value orientation resulting from internalized social or political norms during the process of socialization cannot be called a "constraint." The reason is that the existence of a society presupposes some social norms that are observed by all the members in it. Such norms as there are, are observed in the process of socialization and internalized as an individual's value orientation, forming part of that individual's character. Thus an individual in a society will always be constrained by corresponding moral norms. Can we then say that everybody's behavior is constrained and there is no individual autonomy?

This is the crux of the matter. Some neo-Confucians and their sympathizers maintain that Chinese culture, especially Confucianism and Taoism, encourages the development of individual autonomy.[16] Their explanation makes plain the differences between traditional Chinese views and modern Western views on the notion of individual autonomy.

Traditional Confucian thinking lays much emphasis on the cultivation of internal, transcendental individual autonomy. Confucian scholars believe that humans can, through persistent practice of introspective contemplation and self-cultivation, learn to control all kinds of desires and attain "benevolence" or "sainthood." This formulation shows an important difference between Confucianism and Western philosophical and religious traditions. To the Confucians, the process of immanent transcendence and self-cultivation does not require the help of "God" as emphasized in Western religions. Instead, it relies on one's inner self, emphasizing instead autonomy in the development of the moral sphere. It is different from the modern Western notion of a developed autonomy within individualism. Relatively speaking, Western autonomy emphasizes individual decision in the areas of material choices and personal desires. In contrast, the development of autonomy for the achievement of moral excellence naturally pays little heed to the satisfaction of material needs and human desires. Confucians focus their attention on how to develop and cultivate the moral self, and concentrate on the steps in the cultivation of positive transcendentalism; they neglect the very big influence and constraint exercised on humans by material desires. It is true that the Confucian doctrines of the Song and Ming moralists deal with the conflict between the "reason of Heaven" and "human desires," but the analysis of the negative side of human nature was not thorough enough.[17]

To approach the development of individual autonomy from its moral dimension, as the Confucians would have it, we can see a construction of social norms based on the development of the subjective moral self. This can be understood, to some extent, as developing an individual moral self, and taking that as the center, extending it outward from the individual to the family, then to relatives and friends, weaving together a social relations network based on family and lineage ties. Following the Confucian explanatory model, the autonomy of the moral self is the starting point, and the principles of human relationships that govern the family and lineage networks are developed from it. From an affirmative perspective, since this kind of human relations network emanates from the development of an individual immanent transcendence, it not only makes social transactions possible, but also provides the individual with a sense of "ultimate concern."[18]

This way of looking at the Confucian doctrine as well as its influence on Chinese family and lineage structure and Chinese culture is from a positive angle. If we switch to another angle, and view Confucianism against the backdrop of China's modernization process, we might well see a distinctly non-rosy picture. We could say that the individual transcendental "autonomy" emphasized by Confucians in reality has stifled the development of individual autonomy.[19]

The immanent transcendence emphasized by Confucians unquestionably assumes to some extent that humans can achieve inner freedom by their own efforts. This kind of "freedom" is attained in accordance with a certain set of strictures and norms. These strictures and norms are the *li,* or rules of propriety, that Confucians particularly stress. The extension of relationships from the self to the family and the lineage follows the rules of propriety. In particular, the "five human relationships" exert a normative influence on the individual, the relationship between individuals, and the structure of the family and the lineage. From this, we can appreciate the influence of familism and familial culture on a person. For the Chinese people at large, this influence definitely serves the function of sustaining relationships and allowing them to enjoy peace and tranquility both physically and spiritually. But they could scarcely be considered to have attained the ideal of autonomy through an immanent transcendence. It could be that such an ideal has been transmuted to become an attitude variously manifested as "happy to lead a simple and virtuous life," "contentment brings happiness," or "contented with their lot." But no trace of autonomy might have occurred in their minds.

Over the 2,000 or so years of Chinese familism, other than its function of sustaining relationships as noted above, we can see that it has also constrained people's thinking and behavior. The five human relationships have become relationships that revolve around what Hsu calls the "father–son dyad."[20] And they have become a model of a following-the-one-senior-to-you kind of obedience to the rules of propriety. The model is one in which the father–son relationship subsumes all other social relationships, especially that between the two sexes.[21]

From this perspective, as noted above, we do not see that in the course of daily life ordinary Chinese people are in any condition that can be described as exhibiting the autonomy of an immanent transcendence. What we see instead is the constraint imposed on individuals by Confucian morality, whether in their outward behavior or in their spiritual development. This aptly shows how autonomy relates to the constraint from the internalization of social norms mentioned previously.

We shall not elaborate on this topic now, but need only point out that with reference to the development of modern Chinese society, the constraint coming from familism lies at the very heart of the problem. Familism is often regarded as one of the main factors contributing to the suffering of the modern Chinese people. But it should be cautioned that this judgment is made not in an absolute and universal sense. It is not intended to condemn the goal of immanent transcendence and familism for having negative effects on the behavior of the Chinese. Rather, the Confucian version of autonomy and the effect of familism are examined in relation to the development of China's modernization and with it the modern conceptions of human existence. In this regard, we put forward the proposition that the Western concept of autonomy emphasizes more material choices. With this comparative perspective, we shall try to understand the behavior of the Chinese, particularly under the influence of a market economy after 1979. In the following pages, we shall follow this line of thought to portray the social relationships and the development of autonomy under the new market economy.

In recent Western political thought, the relationship between market economy on the one hand and freedom or autonomy on the other is a matter of scholarly concern.[22] Ideally, the market economy provides a social matrix in which there is fair competition so that individuals can make the best use of their talents and their resources; it is in this way that it influences the social structure and individuals. A market economy is different from the planned economy that was in practice in China after 1949. The main difference is this: A market economy emphasizes that the optimal use of resources depends on the rational decision of each individual. Clearly, this presumes the existence of private ownership. It also stresses that the reward a person gets is determined by the effort put in by that individual. Market economy does not posit a clearly defined concrete collective goal to direct the effective production and use of resources. There are only individual goals that drive one to generate and use resources. From this, we can see that the market economy involves certain value orientations besides the efficient use of resources. One of them is respect for individual goals, that is, respect for individual decisions.

This way of developing autonomy is very different from the Confucian concept of autonomy. The autonomy of Western society is very much tied to personal desires. If we admit that while individuals identify with the "larger self" of society they also need to allow personal desires to develop to a certain extent, then we can see that under familism and the Confucian doctrine, the growth of personal desires is considerably suppressed. Let us suspend judgment for the time being, in order to avoid the question of good or bad. Inherent in the development of modern society is the creation and strengthening of personal desires, especially material desires. The concept of the individual produced by familism or Confucianism and that produced by modern society are to some extent in conflict. The two theoretical frameworks of Western autonomy and familial culture are thus suitable tools with which to examine the obstacles encountered in the development of individual autonomy during the process of China's modernization.

The brief analysis above adopts a positive stance to analyze the relationship between market economy and some value orientations. As intimated already, the Western concept of autonomy is more concerned with making personal choices at the material level than the spiritual. This orientation is clear in a market economy. A market economy not only respects personal decisions, it encourages an individual to employ the most efficient means to reach one's goals or to satisfy one's desires. Simply put, to create wealth and get rich are valued positively in a market economy. This is the distinguishing mark of capitalism.

We will not go deeper into such questions. It will suffice to point out that this characteristic of capitalism was discussed in great detail in Western classical sociological works. In particular, Karl Marx, Emile Durkheim, and Max Weber all made in-depth analyses and criticisms of it.[23] To sum up their views briefly, these thinkers are aware that the development of a market economy will lead eventually to people using the market mechanism and its related values to judge all those things that do not belong to the market and even all human activities. Human relationships then become commodified. Thus, they all point out that compared to the past, capitalistic society can satisfy human material desires and enables human "rational" powers to grow to a greater extent. At the same time, these scholars demonstrate from different angles and levels how lost human beings are living in a capitalist society. The neo-Marxists of the mid-twentieth century, especially those constituting the Frankfurt School, show how market-driven capitalist society controls people through "instrumental rationality" or "purposive rationality."[24] Instrumental rationality means the achievement of one's goal by the most efficient resource and means. The advances in science and technology of the last two centuries have invested instrumental rationality with an attractive aura.

Viewed from another angle, market economy encourages the pursuit of wealth, and thus becomes the dominant value in society. Add to these the excessive emphasis on instrumental rationality and the relationship between humans becomes a relationship of means and ends. The "ultimate value" and "ultimate concern" of human existence are replaced by money and power. Simply put, as pointed out above, the market mechanism is intended to determine the price of commodities. Now it has become the main criterion in every facet of human life. In this way, every human affair, behavior, and activity are commodified. This is a diagnosis of contemporary society proffered by quite a number of contemporary Western scholars.[25]

In a market economy, we see that people strengthen the development of individual autonomy through the control and exercise of their own wealth. Concurrently, the value system of individual determination and the idea of fair exchange are attested. However, the market economy as developed until now has exerted a different kind of pressure on people. If we look at the development of China after 1949, we will see that 1979 is a watershed. The reform and opening-up policy after that time has the creation of wealth as one of its main goals. The aim is to enable some people to become rich first, and then to have a trickle-down effect on people in other provinces. This process is based on the operation of market economy. However, in light of the analysis above, market economy also confers upon the Chinese people a completely different set of values, prominent among which is the development of individual autonomy.

When we set the changes against a backdrop of familism and post-1949 communism, we can see the nature of the changes more clearly and more starkly. In the stories told in this book, we see how the people, through the pursuit of wealth, make all kinds of decisions in their daily lives to further their own interests and in so doing develop their autonomy. On the other hand, market economy encourages the acquisition of profit, which in turn fosters the value system of capitalism. This value system turns on the people who believe in it and controls them. If we combine this profit-first value system with the weakening influence of familism, we can both see and explain more clearly how the "dissolution of social order" comes about.

The Critical-Hermeneutic Perspective and Indigenized Research

The arguments presented so far might seem to suggest that we hold a rather dim view of the role Chinese familism plays in modernization, while taking a positive view of feminism and the development of autonomy under a market economy. Some people would say that such a stand raises two methodological issues in social inquiry. First, will the involvement of the authors' values in the study interfere with its neutrality? Second, will the employment of Western theories to study and criticize China's conditions mean placing Chinese phenomena in a Western straitjacket? The latter also concerns the problem of indigenization or sinicization in social inquiry.

Regarding the first question, the authors of this book do not believe there is such a thing as value-free social research. Social research will inevitably carry the values of the researchers and the existential meaning of their social and cultural contexts. We believe that without these presuppositions, it is impossible to understand and study any social phenomena, because social phenomena only become accessible in their concomitant cultural contexts. In fact, since the 1970s, the idea that social research is value-laden has been a common understanding among nonpositivist researchers. They differ only in the different views held in regard to the degree of value involvement.[26]

That is to say, in the collection, presentation, and analysis of field data, it is impossible for the researchers to withdraw their own set of values. To put it another way, the understanding of the phenomena and the interpretation of them cannot be sharply separated. But then the question at issue will be: Which approach to understanding and evaluation can avoid straitjacketing China's affairs? To answer this question, we cannot appeal just to discussion at the theoretical level. Rather, we have also to take into account specific conditions, that is, the specific conditions of our research.

With respect to the angle and scope of our research, we affirm some Western theories and viewpoints and to some extent harbor a negative view of familism. In the collection of field data, though, we try to understand the people's thinking and feelings, the happenings in the villages, and their historical development in the native context of the Chinese village. This is the viewpoint of the method of participant observation. It emphasizes the understanding of social events from the actor's perspective. Some concepts regarding human relationship in familism are an appropriate point of entry for understanding individuals in the villages. This approach has been called the phenomenological or interpretive social research method.[27]

However, the authors of this book do not just conduct the research from this angle. As pointed out in the above discussion, any social research will involve the researcher's standpoint. This includes the social condition in which the researchers find themselves, the cultural context, and the kinds of questions with which the researchers are concerned. In this study, since we are Chinese, to a large extent we can identify with the cultural feelings of the people we study. And since we live in Hong Kong, we naturally view things from our own social context, that is, closer to contemporary Western society. The concepts and angles we employ in our research are rooted in such contexts.[28]

In other words, what we are practicing is meant to manifest the research stance and methodology of critical hermeneutics.[29] According to this view, any kind of social research and understanding of social phenomena is to be understood as a dialogical process in which fusion of the horizons of the researcher and the researched takes place. The researchers take an empathetic view of the situation of the researched, but at the same time, will pass judgment from their viewpoint on the viewpoint of the researched. In this process, the researchers will come to understand more the viewpoint of the researched. This understanding in turn enables the researchers to reflect on their own viewpoint. This is a dialogical and evolving process. It is this cooperative process that makes understanding possible.

To sum up, we use familism as a point of entry for achieving understanding, a means of making sense of the issues and ideas about marriage and relationship between the sexes among the Chinese people. On the other hand, the Pearl River Delta is undergoing cataclysmic changes in the process of modernization, and the issues of marriage and sexual relationship there have been made problematic at the cultural, social, and individual levels. Such issues not only lead us to understand the ideas about marriage and relationship between the sexes in contemporary China, they also lead us to assess whether traditional Chinese concepts of marriage and the relationship between the sexes under familism are reasonable. In all this, Western theories provide interpretations of a different way of life with which one can compare the Chinese situation. This comparison not only throws into sharp relief the characteristics of the Chinese way of living but also examines the "legitimacy" of such a way of life. This approach combines understanding and critical appraisal as well as Western and indigenous angles. Chapter 14 of this book is a reflection on methodology, where we will further explain our point of view.

2

Baixiu Village

Past and Present

Baixiu Village is located east of Qingyang Town in the Pearl River Delta and is under Qingyang Town's administration. The distance between them is about six kilometers. Before 1993, the main road linking Baixiu Village and Qingyang Town was a small two-way road, one lane on each side. By 1995, the road on each side had been widened to become a three-lane highway.

The area of Baixiu Village is about five square kilometers. The residential area takes up about one square kilometer and the rest is farmland. Since the economic reform in 1979, about two-thirds of the arable land has been filled to make way for factories and industrial development.

The population of Baixiu Village is about 2,000. Most of the people have Wang as their surname. According to the Wang genealogical table, the forebears of the Wangs migrated to Guangdong from Shandong in the waning days of the Southern Song dynasty. There have been more than twenty generations of Wangs over the more than 800 years since the first settlers. There are over ten other surnames in addition to the predominant surname, Wang. Since the early 1980s, there has been a steady stream of outside investors from Hong Kong, Taiwan, and other places who set up manufacturing plants there. From 1993 to the end of 1994, the number of factories increased from twenty to over forty, and the number of outside workers living in the dormitories of the factories tripled, from about 10,000 people to over 30,000, about 70 percent of them female. They come mainly from Guangxi, Hunan, Hubei, and Sichuan provinces.

The workers' houses are located on the north of the village and are mainly old-style farm houses, some derelict and seedy. After the reform and opening-up policy, a portion of the villagers became rich, and a few modern two- and three-story buildings appeared. In the absence of city planning, new buildings and old houses exist cheek by jowl, and the overall impression is one of confusion.

In the heart of the residential area lies the market. Early each morning, the market is a beehive of activities. Merchants display their vegetables and meats in stalls, but they do business only in the morning. In the afternoon and evening, only grocers with vegetables, which are spread out on a thin sheet on the ground, are at the market.

Before 1995, the factories in the village were situated in the industrial districts in the northeastern and southern parts of the residential area. After 1996, the village management committee expanded the industrial area, and moved all the factories that stood near the residents' houses to the south industrial area. Since there are large numbers of outside workers, many shops and restaurants sprang up in the village, along with a small shopping mall where the workers could buy daily necessities. After some city planning in 1996, the industrial area has become an industrial community.

As a large amount of farmland has been turned over for industrial use, only one-third of the original farmland remains. Even there, some farmland has been turned into fish ponds. What is left of the farmland has been rented to farmers from other provinces. The villagers as a whole do not earn their livelihood by farming. Baixiu Village has thus begun to shed its traditional farming character.

The changes that have taken place in Baixiu Village are representative of similar changes around the Pearl River Delta after the reform and opening-up. That is why we chose this village to be the target and location of our fieldwork. In the spring of 1993, our research team of five people began the work of gathering field data. To facilitate communication with the villagers, we rented the ground floor of a three-story building close to the village market. There we set up our fieldwork "research center." Our research work lasted more than three years, with the main bulk of the research data collected in 1994 and 1995. In those two years, we made more frequent trips to Baixiu Village than in other years, and our research came to an end by late 1996.[1]

Baixiu Village After 1949

After the Communists took over China in 1949, Baixiu Village underwent unprecedented changes just like any other Chinese village. In 1951, the government began the land reform movement. Land reform distributed land on the principle that the tiller of the soil would get land to work on. The government took land away from the landlords and redistributed it to landless farmers so that every family in the village had a piece of farmland of its own. This reform did not affect the renters of land much, since they could still ply their trade. The difference was simply to transfer their rent based on yield from a landlord to the state. The reform did not affect small landlords much either. They had not had much surplus land to rent out anyway. After the land reform, they, too, continued to ply their trade, only to transfer part of the harvest to the state as rent. The sector most affected by this movement was the rich landlords. They could not own land anymore, and those who relied on collecting rents now had to work in the paddy fields.

Three years after land reform, Baixiu Village began to organize mutual-aid cooperative units. Villagers voluntarily formed cooperatives. After families with many working hands or a strong labor force finished their day's work, they would be encouraged by the government to help those families with insufficient labor to finish their work as well.

After 1954, Baixiu Village began to experiment with lower-level agricultural cooperatives. During the land reform years, the villagers would give up part of their harvest to meet government procurement quotas and then they could either sell the rest of the harvest to the state or keep it for their own use. The rise of lower-level agricultural cooperatives signaled the start of the road to collectivization. Whatever the land yielded was distributed by the state. The distribution was done according to the number of workers in each household who went to the field to work and the area of land each worker cultivated.

Just one year later, the pace of collectivization of the country quickened. Baixiu Village kept up with the pace and advanced to a higher-level agricultural cooperative. The principle of distributing earnings to a household in proportion to the total area its members cultivated was abandoned. Instead, each peasant received earnings according to how much he or she worked and according to labor performed. The work-point system was initiated at that time. In 1958, people's communes were set up all over the country. Qingyang Town followed other counties and established a people's commune. Baixiu Village became one of the production brigades of the Qingyang People's Commune. The Baixiu Production Brigade was divided into six production teams. Each person's daily work was rated by the production brigade in terms of work points. After a year, the production brigade converted the points gained into wages. Each person on average made about 300 yuan a year. But this sum was not the take-home pay. Each person got thirty-five catties of basic grain provisions, and this had to be deducted from the work points. If a household had four people and only one laborer, then after deducting four basic grain provisions, that household might not have any income, or worse, be in the red.

From the establishment of the people's communes in 1958 onward, Baixiu Production Brigade followed the country's collectivization, "eating from the big wok." Villagers went to the village canteen to have their meals. In order to raise the amount of farming produce, the brigade followed the method of cultivation recommended by the government. This eventually led to reduced production, to the extent that there was not enough grain to share. Older villagers still remember that in 1960, not only did farming fail as a result of the Great Leap Forward launched in 1958, but they had to give additional grain to the government because China had to pay back her debt to the USSR. The result was that Baixiu Village, like so many other villages, suffered famine. In 1962, Guangdong Province saw a massive exodus of illegal migrants to Hong Kong due to famine. Many members of the Baixiu Production Brigade escaped to Hong Kong at that time. When the Cultural Revolution began in 1966, Baixiu Village entered its bleakest years, and it was not until 1978, when economic reform and the opening-up policy got under way, that the livelihood of the village improved.

In retrospect, although the land reform movement after 1949 turned the ownership of the land over to the state, it did not much affect the economic activities of ordinary peasants, but did affect rich landowning farmers. Every village household had to pay tax in grain; what was left was theirs. It was not all that different from the years before 1949.

A radical change occurred after the farming cooperatives were converted into people's communes—a big step in collectivization. All that the land yielded belonged to the production team. No villagers owned any land that they could manage and cultivate on their own. All livestock belonged to the team and were fed by the team. Personal ownership of such animals would be branded as capitalist. The villagers' incomes were calculated in the work-point system after being assessed by the team. The Baixiu Production Brigade was like a big factory. Villagers got their wages based on their work performance. The only difference between a factory and the commune was that in a factory, wages are distributed to each worker, but in the Baixiu Production Brigade, all the wages of the members of each household were handed over to the householder, that is, the head of the household. The more members that a household had and the more they worked, the higher the wages would be and total household income would be that much higher. Besides, the recipient of the wages was the householder, so the head of the family had a strong hold on the economic resources of the family. On these two counts, collectivization did not turn the economic structure of the family around.

After 1949, the collectivization of economic activities was a notable national development; another was that the government waged a campaign against "feudal culture." Traditional folk religious activities such as *feng shui* (geomancy) and divination were banned. During the decade after 1949, villagers could still buy joss sticks and candles to burn in altars put up at home to honor ancestors, but in the latter years of the 1960s during the Cultural Revolution, even lineage activities were strictly prohibited. During the spring and autumn times of ancestor worship, villagers could only go to their ancestors' burial place surreptitiously to sweep the tombs and pay their respects; this was done with the least ceremony, in order to avoid attracting unwanted attention.

In the wake of the antifeudal political movement, the Wangs of Baixiu Village lost their genealogical book—it was burned. The chief ancestral hall where all the Wang ancestors' shrines were placed was converted into a meeting place for villagers, and was even demolished in 1957. Some of the twenty-two ancestral branch halls, branches of the Wang lineage, were also torn down. Those not demolished were turned into residences for poor peasants who had only straw huts for living quarters. The government encouraged the poor Wang peasants to criticize and denounce their more well-to-do relatives. The latent economic conflicts that had existed between different branches of the Wang lineage were intensified. The more affluent Wang members were denounced as landlords or local tyrants. Some people of an older generation recalled that one branch of the Wangs, the descendants of the Dezu branch, who had played the leading role in pre-1949 days, were criticized and denounced most severely, and their leadership role taken over by other Wangs. For the Wang lineage, then, the "antifeudal" political movement brought about only internecine fighting among the different genealogical branches.

In 1950, the government promulgated the new Marriage Law, with the purpose of eliminating what was regarded as the "feudal" traditional marriage practice. Compared to what was practiced before 1949, the new law encouraged freedom in marriage. Marriage between persons of the same surname and between persons in the same village was allowed. In 1952, a work team came to Baixiu Village to explain the concepts behind the new Marriage Law and urge people to select partners on their own. Despite its promulgation in 1950, from the 1950s through the 1970s, the villagers in Baixiu Village still approached a matchmaker to find a spouse for their sons and daughters. The villagers pointed out that there had never been a marriage between persons of the same surname before 1949, and the first such case took place only in 1960. Thereafter, such cases were few, but increased in number in the 1980s.

In the early 1950s, the Chinese government formed the Women's Federation and sent some work teams to Baixiu Village to launch a women's movement. Women's groups were formed in the village to teach women to resist unreasonable traditional customs. One example given was that if the mother-in-law was too strict, the daughter-in-law could resist such harsh treatment and bring a complaint to the village government. The women's group also taught women to be self-reliant, earn a living, and not depend on others. If a couple had the means, they should leave the home of the husband's parents. If the couple did not have the means, they could at least try to set up stores separate from that of their parents. The wife would try her best not to be enslaved by her father-in-law and mother-in-law. The upshot of this movement did not radically change the status of women in the village. However, when older women recalled those days of reform, they felt that the women's movement enhanced their status and rights compared to those before 1949.

Economic Reform and Opening Up

The policy of collectivized economy came to an end only with the death of Mao Zedong in 1976. After 1978, the people's commune existed only in name. The work-point system of the production team was abolished. A new policy of allotting land to each household took its place. Each villager in Baixiu Village received two *mu* (1 *mu* equals approximately 0.067 hectare) of farmland for cultivation. The state exacted 100 kilograms of grain for each *mu* of rice field as tax. Each household could raise its own chickens, geese, ducks, pigs, and so forth. The policy of allotting land to each household was similar to the land reform of the 1950s. After 1979, villagers could work outside their village. The condition was that each worker had to give 20 yuan to the work team. After 1980, collectivization of the village economy broke down completely. Villagers could look for work where they wanted, and payment to the brigade was discontinued.

At the end of 1978, China embarked on the "economic reform and opening-up policy." Some Hong Kong manufacturers started to move their factories to the villages in the Pearl River Delta. In the last months of that year, the first Hong Kong factory that manufactured plastic products with imported materials moved into Baixiu Village. In 1979, a Hong Kong knitting factory moved in. The villagers made monthly wages of 70 to 90 yuan, which increased to 120 yuan after three months. As the villagers could earn far better wages than from farming, they rushed to get factory jobs.

By 1988, the number of factories set up in Baixiu Village increased to eight, and the demand for workers increased proportionally. Outside workers from other provinces started to make their appearance. They totaled about 500. Their wages were equivalent to those earned by the locals, but the locals enjoyed much better chances of promotion.

With the increase in factories, the village government started to requisition farmland from the production teams to build factories on. About three hundred *mu* of farmland were bought back, at the price over one hundred yuan or so per *mu*. In 1991, the requisition effort was at a peak: nearly half the farmland was requisitioned for industrial use. In the same year, the government stopped releasing the prices of the transactions, and each household would get its share of the money from the land sale only when the manufacturers paid rents to the production team for the factory land. In 1992, of the six production teams in Baixiu Village, the third and sixth teams had no more farmland to cultivate because most of it had been turned into factory sites and what was left was converted into fish ponds. By 1996, only one-third of the original farmland was left in Baixiu Village.

During the process of the requisition of farmland in 1991, because the village government did not give back the profit made in the land sale or publicize the prices of the transactions, the villagers became very agitated. In early 1992, some villagers prepared banners for a public demonstration march through the village. The government used every forceful and persuasive means to calm the villagers down. This sort of discontent was also seen in villages near Qingyang Town.

In 1991, the village government converted the south part of the village into a factory zone. In 1993, when we began our fieldwork in Baixiu Village, there were about ten factories together with ten or so small shops selling clothes and daily necessities in the industrial area. In addition, there were three or so restaurants and a small cinema. In the evening, the shopping district was crowded with people. Workers thronged the streets and shops. Because entertainment establishments were few, the front lobby of the cinema was always packed with people just before a show.

The number of factories in Baixiu Village increased to twenty by 1993, the number of outside workers to more than 10,000. Construction work by the government continued near the southern factory area, steadily expanding its size. Between 1993 and 1994, the farmland adjacent to the factory area had become a large construction site, and factory after factory sprang up. By 1996, the factory zone was several times its original size. At the same time, the government moved all the factories that were built in the northeastern part of the residential area to the factory area. The expanded factory area is served by broad cement roads that carry container and other trucks. Along the roadways are shops of all kinds that sell a large variety of everyday goods. A new and bigger cinema was built to meet the entertainment needs of the workers. Viewed in its entirety, the factory zone had become an industrial complex.

The factory area had been expanded and its planning improved, yet the wages of the outside workers did not catch up. Some migrant workers have better educational qualifications and ability, but their promotion prospects are far less than those of the locals. The senior staff of a factory consists almost entirely of local people. By and large, locals receive priority treatment. The wages of a novice factory worker in 1996 were no different from those in 1993—about 300 yuan a month. Any wage increase has been offset by the rise in cost of living in the Pearl River Delta during those three years. To cite a simple example, the cost of a plate of rice and meat increased from 3 yuan to 5 yuan in three years.

Compared to outside workers, the village workers are treated more generously. When they first join a factory, they may take up a low-level post. After a few months, they might be promoted to head a production line or a junior supervisory position with the job of overseeing the outside workers. Among locals, men have a greater likelihood of taking top positions than do women. The general managers of the factories in Baixiu Village are almost all men; women can aim for supervisory positions only at the lower ranks.

Since the rural economic reform, villagers have not been limited only to factory work; they can engage in various kinds of business. Some go into fish farming, some to pig raising, and others to fruit cultivation. The industrial development of Baixiu Village brought many opportunities for investment. Filling farmland, building factories, and expanding the factory zone brought on the construction industry. The influx of large number of outside workers meant many opportunities for small business. Many villagers opened grocery stores, clothing shops, and restaurants in the factory zone or in the village.

Economic development in Baixiu Village has been brisk. However, investment opportunities are firmly in the hands of the secretary and party members of the village government. For example, some party members have a monopoly on the catering services for the factories' canteens. Others award construction contracts for filling farmland and building factories to companies they own. Even though villagers might have enough capital, they still have to tap their social network and special relationships if they are to secure profitable business contracts. In Baixiu Village, the nouveau riche are usually party members of the village government or villagers who are close friends of these members.

The quality of life in Baixiu Village was poor after the setting up of the people's communes, and it was not until the period of rural economic reform that villagers began to enjoy a better life. Before 1978, only a few households possessed an electric rice cooker; now almost every household has the basic electric appliances. Some rich party members have lavishly furnished homes and two to three expensive European-made cars.

The economic reform and opening-up policy had more than an economic impact on fostering free economic activities; it also revived traditional lineage activities. After 1979, with economic prosperity, the political environment became more relaxed. Ancestor worship and the ritual offering of sacrifices in spring and autumn reappeared. In the early 1980s, some branches of the Wang lineage gathered all the male descendants to attend sacrificial rites in spring and autumn. Even those men who had migrated illegally to Hong Kong came back for the occasion. In addition to worshipping the forebears of their own branch, the Wang lineage had all the men worship at the tomb of the first Wang progenitor.

Around 1984, a group of older-generation Wang villagers talked about rebuilding the demolished main ancestral hall of the Wangs. The idea received the support of the Wang descendants in Hong Kong. In 1985, the plan to rebuild the hall was launched, and in the short span of a few months, a sum of 180,000 yuan was raised by the descendants in Hong Kong. Baixiu villagers contributed 20,000 yuan. In the preparatory stage, the secretary of the village government opposed the idea for a while, but since other villagers in the county were rebuilding ancestral halls, the project went ahead. Finally, the ground-breaking ceremony took place in 1986 and the hall was completed in 1987. On the day of the opening ceremony, all the Wangs who had gone to settle in other villagers in the same county gathered in the village to offer greetings.

After the ancestral hall had been built, another project was broached—to compile the genealogical book of the Wangs, which had been burned during the Cultural Revolution. Since the list had been destroyed, the villagers could only gather together and collate those lists that had been hidden or preserved by kinsmen who escaped to Hong Kong. In addition, they interviewed Wang households and recorded information about the new generations of males since 1949. Finally, in 1993, they completed the task of re-creating the Wang genealogical book.

Subsequent to these two events, the elders of the village proposed to build a Wang Fraternity Association, ostensibly for the recreational use of the Wangs who come back from Hong Kong to visit from time to time and of the villagers in general. This proposal was met with some reluctance from the secretary of Baixiu Village, who probably felt that the Wang lineage was getting too powerful. He suggested that the club be called "Baixiu Village Fraternity Association." Although the club was not a lineage association, the first board of directors was, without exception, all Wangs. Most of them had been active in rebuilding the Wangs's main ancestral hall.

Although lineage activities were revived, only a small number of young people joined them. Most young people were not interested in such activities.

Among those elders who participated regularly, there were some who had their eyes on leadership positions and this created periodic squabbles among them. Some elders who were enthusiastic about lineage affairs were somewhat discouraged.

Toward the end of the 1980s, with the economy booming, Qingyang Town saw the beginning of the sex trade, which spread even to the rural areas. In the 1990s, the sex trade mushroomed like bamboo shoots after a rain. It took place in hair salons, karaoke nightclubs, cafes, saunas, and massage parlors. Since the road from Qingyang Town to Baixiu Village was expanded, the number of karaoke nightclubs increased from six or seven to over twenty. At first these places of sexual gratification were for merchants and technicians from outside China, but they gradually became frequented by the local men. The more well-off villagers went to karaoke nightclubs and the lower-income villagers went to hair salons for sex services.

The male villagers of Baixiu Village, old and young, have visited such places, almost without exception; some even go five times a week. The "ladies" who provide such services are mostly women from other provinces. At first they came to get jobs as factory workers, but there were too many of them. For instance, in 1994, there were more than 20,000 female workers from other provinces in Baixiu. Some of them probably found the temptation of money and material goods irresistible and left the factories to become "ladies."

After more than a decade of rapid growth, around the end of 1994, the economic development of Baixiu Village began to slow down. Although the village government still built new factories, only one company moved in during 1995. The various trades and businesses entered a quiet period. The chances of becoming an instant millionaire were now fewer, and work was harder to find than before. The people of Baixiu Village were waiting for another wave of reform.

Part I

Finding a Partner, Love, and Marriage

3

Marriage, Familism, and Autonomy

Chapter one states that traditional familial culture emphasizes the "father–son relationship" as the axis of the family. The ultimate value and objective of the family are to continue and extend the blood line of the paternal side. The patriarchal culture engendered by the "father–son relationship" creates the differential mode of association in which the old and the young have their proper place. It defines the relationship between the sexes in the family as one of men dominant, women subservient. Thus, in the traditional marriage system, parents choose spouses for their sons. In doing so, on the one hand, they consider the prospective daughter-in-law's likelihood of continuing the lineage by bearing children, and on the other hand, they set great store by her character. They look for someone who is respectful of the parents-in-law and obedient to the husband. They value such qualities more than the feelings the son might have toward his future spouse. This is because in the eyes of the older generation, marriage is not just a relationship between husband and wife, it is a function that satisfies the ethical dictates of familial culture.

In other words, we can say that from the angle of the traditional familial culture, parents make decisions about their children's marriage not based on the latter's own likes and dislikes but on the ethical ideals of familial culture. Personal feelings and preferences, theirs and their children's, are not important; they are to be secondary to the well-being of the lineage as a whole. In contrast to the values and beliefs of the West in recent times, the traditional

Chinese ideas about marriage are oppressive; they suppress individual autonomy and reinforce the inequality between the sexes.

In 1950, the Chinese government promulgated the People's Republic of China Marriage Law. It attempted to reform what the government considered "feudal" traditional concepts about marriage and to eliminate the unequal and not voluntary traditional marriage system. The government also launched a propaganda campaign to educate women about new concepts of marriage, saying that marriage centers essentially on husband and wife. It criticized the so-called traditional ideals such as "three obediences and four virtues" for women as really nothing but means of oppression.

Despite these efforts in the 1950s and 1960s, there were few cases of people self-selecting a member of the opposite sex for a spouse or going through a process of courtship before getting married. The great majority of marriages were still arranged by the parents.[1] Baixiu Village is no different from other places in this regard. Marriages that took place in this period almost always did so through a matchmaker. Some villagers point out that they did not want to be criticized by their peers and did not actively seek out members of the opposite sex. Some could not bring themselves to oppose what their parents arranged for them. They still hung on to the traditional way, accepting the authority of the parents and observing strictly the dictum that men and women should not get close to each other.[2] Such attitudes bespeak, to a certain extent, the fact that since 1949, marriage is still based largely on familial normative constraints. From the fieldwork in Baixiu Village, the reasons for this can be summarized as follows.

The new marriage law and the movement to educate the masses about marriage might have shaken traditional ideas about marriage to some extent. Whether peasants would then abandon their long-held traditional familial culture instantly and embrace new ideas about marriage is doubtful. Traditional familial culture is a culture that is interdependent on the principles of human relationships in a farming society. If the economic activities centering around farming were to be altered, then familial culture could face a real challenge and might even break down. What happened was quite different. During the period between 1949 and the setting up of the people's communes, although villages underwent a series of land reforms, peasants' economic activities were still largely farming, and the family was still the basic economic unit of the village and the fundamental economic formation had remained unchanged. After the setting up of the people's communes, the production mode based on the family as the unit gave way to a collective mode in which the production brigade was the unit of production. Peasants were employed, as it were, by a big firm or big factory. Their means of livelihood was still the tilling of the soil, and the wages were given to the head of the household. From the perspective of the change, during the period from the land reform to the people's communes, the family could still be regarded as the basic production unit of the village. The householder, as before 1949, still held the economic power of the family.

Moreover, in the process of collectivization of production, the production brigade in a people's commune was made up of villagers. In many villages in Guangdong, this consisted of people from the same lineage. Take Baixiu Village as an example. Although there are people from other lineages with different surnames, most villagers are members of the Wang lineage. After 1949, lineage activities were progressively forbidden, but because the majority or all of a production team came from the same lineage, the relationship holding among lineage members was still influential.[3] In the period between the 1950s and 1970s, the government hoped to replace traditional feudal culture through a series of political movements, yet in villages, political movements often led to infighting among the various branches of the lineage. Consequently, unity within the lineage branch was strengthened and the relationship among related families cemented.

It can be seen from these changes that even when the villages in Guangdong Province underwent political challenges from the 1950s through the 1970s, the influence of traditional familial normative constraints did not disappear. This explains to some extent why even after the implementation of the marriage law, finding a spouse was still largely in the hands of the parents.

Starting from the 1970s, some villagers in Baixiu Village took the first step and sought to choose their own spouses. This might reflect the trend of a stronger sense of autonomy in marital decision making. Nevertheless, in both the process of selecting a partner and in marital life, the villagers are to a certain extent under the influence of traditional family norms. The protagonist in chapter 5, Li Zhichao, has a marriage not arranged by his parents, unlike that of the old man in chapter 4, Wang Zhenqiu, but his marriage and the relationship between husband and wife are to a great extent a replication of the traditional model.

Since 1979, with the implementation of the economic reform and opening-up policies, the traditional agricultural economy has been replaced by the economy of industries or small business. In this district, the quality of life and the autonomy of the use of resources have been greatly enhanced. The political climate is more relaxed compared to that before 1979. Traditional lineage activities that were banned after 1949 blossomed in the 1980s. The rapid formation of a commodity economy brought with it a change in lifestyle. The peasants learned about and imitated Hong Kong's way of life through television and other sources. Their perspectives broadened daily.

Young people lost interest in lineage activities. To some extent, the hold of traditional familial culture on the younger generation has decreased.

In considering marriage, more and more young people seek romantic love; in marriage, young couples have begun to take seriously how they feel toward each other. This change would seem to move the axis of the family from the "father–son relationship" to that of a "husband–wife relationship." The story of Wang Qiming in chapter 6 illustrates this change. If continued, this could be the means to level the status of the two sexes. However, the situation of the Pearl River Delta is such that it does not seem to lead to a situation of equality, in which the two sexes have equal opportunities to find a partner. In a village on the economic upswing, a large number of female migrant workers rush in to look for work. This greatly increases the pool from which males select their partners. Furthermore, because the sex trade is rampant, peasants can easily buy sexual services. Such developments in fact lower the status of women. From the woman's point of view, the goal is to find a man whose earning capacity is better than her own. Since out-of-province male workers fare no better than the local males, the choices facing a woman are not increased with the influx.

Part I will document the marriages of three men from three different generations—Wang Zhenqiu, Li Zhichao, and Wang Qiming—as a means of understanding the changes in concepts about marriage, and thereby showing the changes in familial culture over these past decades. It will also demonstrate the tension between familial concepts on the one hand and autonomy on the other. It will show the dominant–subservient relationship that exists between men and women in marriage and in family life. Since all three villagers chosen as case studies are male, the interpretation of changes in ideas about marriage are from the male viewpoint.

4

The Story of Uncle Qiu

The Dictates of the Parents, the Words of the Matchmaker

After Communist rule began in China, the villagers who now make up the older generation in Baixiu Village underwent huge political and economic changes, yet in their daily lives they still hung on to the traditional familial ideas. Although the government promoted new notions about marriage, the older generation still subscribed to the traditional marriage practice of following "the dictates of the parents, the words of the matchmaker." They still firmly believe that this practice is far superior to the romantic love fancied by the young people of the 1990s. This chapter tells the story of Wang Zhenqiu; it shows how villagers of the older generation believe in and stand by traditional views about marriage.

The Tradition-minded Wang Zhenqiu

I got to know Wang Zhenqiu in his barber shop. I was looking for villagers of an older generation to tell me about the history of the Wang lineage. The owner of the general store in front of our research center told me that Wang Zhenqiu was the right person for my purpose, as Wang was one of the most active in affairs related to the lineage in the village.

Wang's barber shop is only three units away from the research center, on the right. It is a small shop, about eight square meters. When I walked into the shop, Wang was cutting the hair of an old villager; three or four other people of a similar age sat on long benches waiting to be served.

Wang was a little over sixty, on the plump side, dark-skinned, and about 160 centimeters tall. He is generally called Uncle Qiu by everyone. He has a deep scar on the left side of his neck; the mark has been there for as long as he can remember, its origin buried deep in distant times. His left foot limps a little, but when he walks slowly, the limp is not noticeable.

Uncle Qiu is a descendant of the fourth *fang* of the Dezu branch family of the Wang lineage. The eldest son in the family, he has a brother ten years younger. Uncle Qiu is an adopted son. His father was thirty-two and, fearing he might not have a son, he adopted him. He was only four years old then, and his place of origin and name are no longer traceable. Uncle Qiu got married in 1949 and had a son who left to go to Hong Kong with his mother in 1957; they come back to the village once a year to see Wang. When the son grew up, he became a truck driver and commuted between Hong Kong and China. At that time he visited his father more often. Then in 1990, he married a young woman in China and they had two daughters. For decades, Uncle Qiu's family members have not lived with him; he is a lonely man.

Uncle Qiu's barber shop is small and stark. It consists of an old armchair for the customer, two old mirrors in front of it, and a narrow board fixed underneath to hold equipment like scissors, razor, and accessories. Two chairs were placed against a wall, and there was a table in one corner with a set of Chinese chess on it. Clumps of cut hair and cigarette stubs were everywhere; a sense of untidiness pervaded the shop.

Since the economic reform and opening up, old-style barber shops like Uncle Qiu's have been in decline. Uncle Qiu's customers are mainly old villagers and children. A haircut costs about 4 yuan, but young people seldom go there. They prefer the hair salons in the next village or in the town, where they can get a more modern style. Those salons are neater and cleaner, and have air conditioning in the summer. The prices are higher, of course. In Uncle Qiu's eyes, these hair salons lure customers by their outward décor; there are no professional skills there. To their discredit, young people just love appearances, he grumbles every now and then, whereas professionals like him who ply an honest trade are fewer and fewer in number.

The way modern hair salons operate is a source of some of the dissatisfaction Uncle Qiu feels; his attitude toward the changes brought about by the reform and opening up is also negative. The industrialization of Baixiu Village has undoubtedly raised the living standards of its residents, but Uncle Qiu sees those changes as having had many undesirable effects. One of his main criticisms of the open-door policy is that Baixiu Village has lost large areas of farmland. With farmlands gone, if there is an economic recession, the village will suffer. Although Uncle Qiu's views have no following among the young, they reflect similar views held by the majority of the older generation.

Besides worrying about the possibility of a shortage of subsistence crops, Uncle Qiu is very unhappy about the way land is sold. He points out that before the liberation, any land to be sold would first be offered to a relative in the same lineage. Only when no one in the lineage was interested would it be sold to people outside the lineage. Now the rule is that land will be sold to the highest bidder, so it ends up in the possession of merchants who move in from the outside. An old villager who is normally a mild-tempered person also huffed, "This way of doing things is all about money. There is now no distinction between relatives and outsiders, and no regard for the needs of the village. This way of doing business ignores the distinction between our own people and outsiders." Uncle Qiu and the old people think that selling land to the highest bidder will eventually damage the ethical relationship among lineage members.

What upsets Uncle Qiu most about the reform and opening-up policy is that since it began, young people no longer respect the senior people of the older generation, nor do they respect the authority of the head of the family. As a matter of fact, once or twice in the barber shop, I heard his nephews and younger people in the same branch of the lineage call him "Qiu Ji" (Craftsman Qiu), or "Head-shaver Qiu." When they left, Uncle Qiu would say to me, "The youngsters now are good for nothing. They cannot tell who is senior and who is junior. All because we have lost our farmland." He feels that since he and others do not farm any more, the economic power is not in the hands of the householders any more, so young people do not respect the elder or care as much for propriety as they did before.

"These young men—they have their own jobs, they make money. Will they listen to their elders? Lucky if you do not get a telling from them. You see how serious the consequence is without farmland! In the past, when we all farmed, the parents had economic power. Were there any children who dared to turn a deaf ear? Now they are rich, free, and beyond the control of their fathers. When we were small, we were respectful of our seniors. But now the world is different." The fact that Uncle Qiu does not like the social changes that the rural economic reform has brought is an indication, to some degree, of his stubborn hold on the traditional way of life.

Uncle Qiu's Marriage

Uncle Qiu began as a barber in 1947, when he was sent by his father to Jinglin Town to be a barber's apprentice. He was nineteen then. He promised his teacher that he would be an apprentice for three years before returning to Baixiu Village. However, in 1949, on the tenth month of the lunar calendar, he received a letter from his father urging him to return to Baixiu Village; something in the family required his immediate presence. So he took a few days' leave of absence from his teacher and hurried home. To his surprise, his father had arranged his marriage for him: he was to wed the next day. Although Uncle Qiu had not counted on getting married so early in life, a father's will had to be obeyed.

Uncle Qiu said, "How could I refuse? Everything had been arranged. What else could I do? In those days, how could we disobey our father? Paternal authority was something of the highest order. My father asked me to learn hair cutting, and hair cutting I learned. I can't compare with young people these days." And so Uncle Qiu was wedded, early in October by the lunar calendar in 1949. Three days after the wedding, he returned to Jinglin Town to resume his apprenticeship; he asked for leave of absence every now and then to go home to visit his father and his wife.

Marriage is a stage in life that Uncle Qiu would certainly go through; that was in the plan. Thinking about the marriage, though, Qiu still thinks that his father's decision came too hastily. He said, with a bit of a grumble, "At that time, I was reluctant to get married. I did not have the means to support a family. I was to concentrate on learning a skill in three years, then I could become a barber and make a living. Out of the blue, I was told by my father to get married. I was not prepared for that, not psychologically. Even when I could become a barber, life might not be as secure as that of a farmer. A farmer will not starve. As a barber, one could starve if there is no business."

Maybe because Uncle Qiu was an adopted son, he had no share in any farmland. Neither did he have an education; he stopped attending after primary school. In 1947, since he was sent by his father to Jinglin to learn hair cutting, he felt that he would not be well off. He had no land to his name, and as he had not completed his full apprenticeship, he had to be married in shabby circumstances. He lacked a sense of security and confidence; he had to support his wife and children in the future. He felt that a big burden had been placed on him.

In addition to financial burden, there was another source of dissatisfaction. Uncle Qiu did not like the fact that his father had not said a word to him about the whole affair and had not allowed him to meet the future wife beforehand. He did not have an inkling about his bride's looks or character.

In fact, however, Uncle Qiu had known something about his wife. They were brought up in the same family when they were children. His wife, Xiao Fen (Little Fen), was brought home by his father to become a child daughter-in-law at the age of thirteen. Before that, she had lived in Nanmen Village to the south of Baixiu Village. Her family was abjectly poor, and she was sold to Uncle Qiu's father. After living in the new home for two to three years, she escaped. It was only later that they learned she had gone to Hong Kong. Hong Kong did not seem to be much better than Baixiu Village, so she returned to her home in Nanmen.

Uncle Qiu's criticism of his father does not seem entirely reasonable. After all, he lived in the same house with Xiao Fen for two to three years, and the purpose of her presence there was clear. Still, he felt that even though Xiao Fen had lived in his house for a time, he had not seen her since she ran away, and he only learned that Xiao Fen was to be his spouse on their wedding day.

For Uncle Qiu and other men of his generation, it was important to know not only the looks of the future wife but also her character. A male relative of Uncle Qiu once told me how important it was to understand the wife's background and character before marriage. This relative said, "Before proceeding with the matching, the matchmaker will give a brief account of the other person's character. If the parents are satisfied, then the next step to explore the family background will be taken. Good family background produces good offspring. If the family background is questionable, the girl could at best only be so! On the other hand, if the parents are good people, and the mother is someone who knows how to accept criticism well while discharging her duties, the daughter will do likewise. That's why we often try to understand the prospective wife's character through relatives and friends."

As Uncle Qiu reminisced about his life, he occasionally showed admiration for the young men now who are so much more fortunate than he was, for they can freely choose their spouse. Even where the parents get a matchmaker to help, young men and women have the chance to get to know the future spouse or even reject the parents' choice.

Although Uncle Qiu expressed some admiration for the freedom in marriage of the young, he also criticized the new way of courtship. In one conversation with me, he pointed out, "The young people now of course have more freedom, but the so-called 'falling in love,' 'going out' are not dependable, not reliable as means of choosing a spouse. 'Falling in love' will lead only to infatuation. Once you are head over heels in love, you are lost and will not see clearly the other person's character."

He further explained, with firm conviction, "The so-called 'free love' as practiced by young people leads to a superficial understanding of each other. It is not possible to understand the other person's character through it. After marriage, the problems are legion and often lead to quarrels and then divorce. Their so-called 'freedom' is too much freedom."

For the young people of the 1990s in Baixiu Village, marital issues are beginning to be directed by personal feelings. To a large extent, Uncle Qiu's outlook shows him to be bound to the ideas about marriage in the traditional familial culture: Marriage is not just a matter between two persons, and the absence of love poses no problem. What is important is that once a wife is married into the husband's family, she should be able to fulfill a range of duties, such as to look after the house, care for her husband, serve the father- and mother-in-law with devotion, and bear children so that the lineage is continued. This view of marriage places the good of the family as the number one priority and downplays the importance of personal feelings. Uncle Qiu is not without feelings, and so occasionally laments his own lack of freedom in marriage. However, he firmly believes in traditional familial concepts, and in weighing the importance of family good against personal feelings, he is on the side of the family.

A Family Life That Is More Apart Than Together

One day, I was talking with Uncle Qiu about some lineage matters in Jiping Village. Jiping Village is a village formed by some villagers who moved out of Baixiu Village. He said that he had wanted to go to Jiping Village to visit his kinsmen; regrettably, he had some family affairs that had to be attended to and could not make it. At the time, I was not yet on familiar terms with Uncle Qiu, and I knew little about his family affairs. He had not told me much about his troubles. Later on, some villagers told me that Uncle Qiu's wife was in Hong Kong. She used to come back to visit him only during the Chinese New Year, but that was some time ago. For the past four to five years, she had not come back to Baixiu. Only his son came back occasionally to see him.

Generally speaking, the elders of the village seldom talk about their wives. Even when I paid Uncle Qiu a visit, I knew very little about what had happened between him and his wife. After quite a long time, we got to know each other better. Eventually, one evening, Uncle Qiu agreed to give me an account of his relationship with this wife in our research center.

Uncle Qiu returned to Jinglin Town a few days after his wedding had taken place, determined to finish his last year of apprenticeship. Three months later, he received a letter from his wife saying that she wanted to go to Hong Kong to work. At this point, Uncle Qiu grumbled, "We have only been married for three months, and together for fewer than ten days. Once she made up her mind to go, she went! I would have let her go after I had finished my apprenticeship, but she insisted on leaving." As these words rolled out of his mouth, Uncle Qiu became visibly upset and he started to stutter.

"Her mind was fixed on going. What could I do? Maybe she learned after marriage that I did not have any farmland, and that I was only a barber's apprentice. She might think that the family would never have a comfortable life. A poor man will get into a situation like that."

Shortly after Qiu Sao (Auntie Qiu, as people called her) left Baixiu Village, Uncle Qiu completed his apprenticeship and went back to Baixiu Village as a barber. In the land reform movement of 1952, even poor peasants got a land share, but because Uncle Qiu was a barber, he did not get any. In 1953, his wife came back to Baixiu Village; life in Hong Kong was just too hard. In 1954, she bore him a son and stayed in Baixiu Village for four years. "Those were bitter years. We quarreled constantly. But looking back now, those were probably the happiest days in my life!" In 1954, Baixiu Village followed the policy of the government and embarked on the collective economic system. Uncle Qiu was assigned to his own job, that of being a barber, while Qiu Sao was sent to work in the fields. When the village practiced collective farming, the villagers worked their hardest but the share of crops each got was not much. One year later, the situation worsened.

In 1957, Qiu Sao finally decided to try her luck in Hong Kong again. "She told me she would go to Hong Kong with our son to find a job. Well, there was no other way. Life here was back-breaking, so I had to let her go. She had gotten a Hong Kong resident's identity card the last time she was there, so it was easy for her to get approval to go." Uncle Qiu's demeanor turned one of sadness as he remembered the year when his wife decided to leave him a second time. He sighed and shook his head. He cursed himself for being useless. This time, he did not display the anger he showed last time.

After Qiu Sao left Baixiu Village, she sent letters to Uncle Qiu periodically. In one letter, she told him that she had her eye on a space under a flight of stairs in a building suitable for a barber shop and she urged Uncle Qiu to go to Hong Kong to make a living as a barber. Uncle Qiu then tried to go to Hong Kong by smuggling, hoping to be reunited with his wife and son.

"I attempted to land in Hong Kong illegally starting December of that year, but my luck ran out. Not once could I succeed in crossing the border. Instead, I was sent to jail a couple of times, and I got to know a few of the prisons in the area. Sometimes I was jailed for one or two days, sometimes for several months. Once, an officer recognized that I had made several attempts. He even scorned me for failing so many times!"

Every Lunar New Year Qiu Sao came back with the son to visit Uncle Qiu. With a job in Hong Kong, she came and went in a hurry every time. Since 1990, Qiu Sao has not come back.

With a cigarette in his hand, Uncle Qiu puffed in silence as the topic turned to his wife who does not come back any more. It touched a raw nerve. He was silent, as if his mind were preoccupied with surveying the past decades that rushed by. After finishing his cigarette, he sat for a while, then left the research center. The next morning, the barber shop did not open for business, and it was only at noon that Uncle Qiu sat alone in his shop puffing away. I asked him why he began his day so late.

"What difference does it make, early or late? The same old thing, and for 8 or 10 yuan. Well, I am old, I am useless, not like you people." He paused, then said, "I didn't go to bed until two o'clock in the morning, and even then couldn't sleep."

"Was it because of what you told me yesterday evening?" I asked.

He did not answer, just nodded. After this interview, Uncle Qiu started to tell me more about his family, more often than not bringing the matter up himself.

Uncle Qiu leads a simple life. Every morning, he goes to the village restaurant to have breakfast, then opens his shop. Business is generally slack. It is only on festive occasions that Uncle Qiu is kept busy. When there are no customers, he sits alone and smokes.

One afternoon, I went to the Fraternity Association to look for Uncle Xing, an old man who is enthusiastic in promoting the lineage affairs of the Wangs, to conduct an interview with him. I saw Uncle Qiu sleeping on a sofa there. It was evening when the interview with Uncle Xing was wrapped up. Uncle Xing and the other elders went home. Uncle Qiu was still sleeping, so I woke him up and asked him if he was going to go home to make a meal. He said, "There's just me to cook for. What's the point? Let's go to a restaurant."

Since Uncle Qiu lives alone, he is of course subject to bouts of loneliness, especially on festive days. Other old men have their children and grandchildren with them, but Uncle Qiu is alone in the house cooking for himself or going to a restaurant alone to eat.

I often had dinner with Uncle Qiu in the village restaurant. The owner of the restaurant has a son and a daughter, the daughter four years old and the son one. Whenever Uncle Qiu goes there for a meal, he likes to play with the son. Sometimes he holds him up and calls his name, as if he were his grandson. One day he said to the owner, "How fortunate it is to have a son!"

As a matter of fact, Uncle Qiu's son married in China and has two daughters. They live in Daixia Town, not far from Baixiu Village. According to the villagers, Uncle Qiu seldom visits them.

I once asked out of curiosity, "Do you go to visit your granddaughters?"

"No. What's the point?" he answered listlessly.

"Because they are granddaughters, you do not go. If they were grandsons, you would go, right?"

"Of course!"

"Did they come to see you this year?" I asked. It took a while before he responded. He shook his head, then said in a huff, "When I die, it doesn't matter whether anyone comes to my funeral. I am used to being alone, all these years."

Uncle Qiu's son did come to visit him in 1995, together with his wife and daughters. That was a year that saw Uncle Qiu become somewhat happier. He told me that when his two granddaughters called him Grandpa, he felt a happiness he could not describe. He always has his son and granddaughters in mind. He understands that his son is preoccupied with work and cannot come back to see him often. However, his daughter-in-law and the two granddaughters live in Daixia Town, only about one hour's bus ride from Baixiu Village, yet they do not come to see him, even during important festivals. He is very upset, yet because he is of a senior generation, he does not want to compromise himself by going to see his granddaughters in Daixia Town.

One could see that Uncle Qiu's lonely life was to an extent the result of his own stubborn character. Qiu Sao left him largely because she could not put up with his bad temper. Also, he insisted on the traditional value of "men dominant, women subservient." This probably spurred Qiu Sao to go to Hong Kong a second time, yet even Uncle Qiu was not aware of his own influence.

An old woman familiar with Qiu Sao pointed out that Qiu Sao's character was actually very compliant. After her baby was born, she looked after him and at the same time went to work in the fields, then in the evening came home to cook and do housework. But Uncle Qiu took the same view of his wife as his mother did, and kept making demands on Qiu Sao to do more for the family as befitting the role of a wife and daughter-in-law. They cared little for her hard work. The old woman told me that Qiu Sao was at her wit's end and could take it no more. And the living conditions in those days were rough, so she eventually decided to leave Baixiu Village and went to Hong Kong to make a living.

What happened to Qiu Sao in her married life is not an isolated incident. In those days, even women married into more well-to-do families would have to take care of many things in the household. If a woman did not have everything done to the satisfaction of the family members, she would be regarded as lacking in good upbringing. If a woman married into a poor household, then her life was even more bitter. Besides doing all the household chores, she had to go down to the fields to work in the daytime. Being such a wife was no better than being a laborer.

According to Uncle Qiu, he and his wife had a bad quarrel in 1990 over a family matter. After that, she left and never returned. I tried to get to the bottom of the argument several times, but was told by Uncle Qiu that he did not understand why his wife never returned to Baixiu. After making inquires among other villagers, I was able to piece the story together.

The cause of Uncle Qiu's quarrel with his wife was the dispute with his brother over the inheritance of a house. Before Uncle Qiu's father died, he had made arrangements for the barber shop Uncle Qiu was working in to be inherited by Uncle Qiu, and another house, a bigger one, to be inherited by his younger brother. The bigger house had room for expansion, but not the barber shop. Seeing that his father favored the blood son, Qiu was inwardly unhappy but did not show it. In 1990, Qiu's father died. His younger brother and his spouse then turned on Uncle Qiu, demanding that he move out of his barber shop and give it to them because as an adopted son, he had no right of inheritance. All the elder and younger relatives said that Uncle Qiu's brother was unreasonable, since according to custom, an adopted son has the right to inherit his father's property. Only children of mothers who remarry do not have the right to come into the stepfather's legacy. Uncle Qiu's brother, however, insisted on reclaiming the barber shop. The elder and younger relatives of the lineage then urged Uncle Qiu to go to court to settle the matter. Qiu Sao also urged him to do so. Uncle Qiu, however, was of the opinion that if the family dispute were taken to court, he and his brother would become enemies. He refused to follow their suggestions.

Qiu Sao was furious with her husband. Clearly, Uncle Qiu's character was such that he would not accept his wife's view but scold her instead. However, when Uncle Qiu mentioned this matter to me, he only said calmly, "Maybe my wife thought I was useless and timid, not daring to go to court even when bullied by my brother. My thinking then was that I should not blow up the matter. Yes, my brother treated me poorly, but I should not bring the law down on him. Ever since that time, she has not come back to see me."

I prodded him further several times, and then he said, "Well, it might be that I was too hot-tempered. But she should not leave just like that. After all, we've been husband and wife for decades!" Although he admitted that his temper was bad, he was never willing to admit that his attitude toward his wife was the reason why she left him.

Uncle Qiu and Lineage Activities

After the policy of reform and opening up was implemented, the government no longer prohibited lineage activities. In the early 1980s, traditional lineage activities were revived in villages in the Pearl River Delta. Like other villages, Baixiu resumed the practice of sacrificial worship twice a year, in spring and autumn. In the mid-1980s, the tasks of rebuilding the ancestral hall and recompiling the lineage's genealogical book were undertaken. Uncle Qiu took part in all these activities. He is now a committee member of the Fraternity Association. By all measures, he is an active participant in lineage activities.

In January of 1995, by the lunar calendar, I joined Uncle Qiu in a lineage activity. The conveners of this meeting were the Wangs of Bailang Village, descendants of the sixth generation of Wangs from Baixiu Village. On a certain day in January, they will call together all the Wangs that hail from Baixiu and those scattered over other places to join in a lineage meeting. Such meetings had been organized by the Bailang Wangs before 1949, but were stopped by the government. In the mid-1980s, after a suspension of thirty years, the Wangs from Bailang reconnected the Wangs everywhere to get them to come again for a yearly meeting. Half a year before the event took place, I got an invitation from Uncle Qiu. In order to better understand traditional lineage activities, I was more than happy to agree to attend.

At about seven o'clock in the morning on the day of the outing, I saw Uncle Qiu coming toward me from a distance. He looked fully rested, in high spirits, and neatly dressed—more so than on ordinary days. He had on a new gray Chinese suit, matching top and trousers, and a pair of new shoes. He approached me, smiling, and said, "Hey, did you have breakfast? It's on me today." I accepted his offer, and accompanied him inside a small restaurant. The place was full of elderly people, all neatly dressed. Even those who were normally in rags had changed into new clothes and brightly polished shoes.

After breakfast, we all boarded the big tour coach parked in front of the village government office, ready to go on the road to Bailang Village. There were about sixty people on the bus, followed by about twenty in a smaller tour bus. The entire party consisted of men who were mostly about the same age as Uncle Qiu. The journey from Baixiu to Bailang took approximately an hour and a half.

I said to Uncle Qiu on the bus, "There are so many people coming to this function!"

"There used to be more! This year is not as good as previous years. In the past we hired two big coaches, which could barely hold everyone. Two years ago, we brought two lion heads to perform the lion dance in Bailang Village. This year we brought only one," Uncle Qiu replied, shaking his head.

An elderly man sitting next to us said, "Nowadays few people learn the lion dance! The young people now don't care for such things. Over the hill, they say! There are so many newfangled things. In the past, our lion dance was superb, now I am afraid we are inferior to other troupe."

As he spoke, there was an expression of wistfulness and helplessness on his face. Then he fell into a long tête-à-tête with Uncle Qiu about the past: how large-scale the meeting was back then, how warmly people had participated. He sighed over the fact that now the villagers were much cooler in their attitude toward lineage activities.

When we reached the outskirts of the Bailang Village, we could see far ahead of us huge banners flapping in the wind. A lion dance was on to welcome visitors. Then there were the sounds of firecrackers and gongs. It was a scene of noise and excitement. When all the representatives from the various villages were gathered together, the two lions of Bailang Village began to move forward, leading the lions of other villages to go into the village. The villagers walked slowly behind. After a few minutes, they reached the door of the local government office. Multicolored banners and streamers were hung, each with the name of a village written on it. Including Baixiu, seven villages took part in this reunion.

A group of old men stood outside the government office to welcome the guests. Upon Uncle Qiu's arrival, several of them came to greet him. It was evident that Uncle Qiu knew the elders of Bailang Village quite well. Since I was carrying a camera, Uncle Qiu was not shy in asking me to take pictures of them. In the hall inside the building was seated a large crowd of seniors from different villages. As Uncle Qiu went in, he greeted the old men from different villages without a moment's rest, as there were so many that he knew. As he talked to them, he withdrew a pack of imported cigarettes from his pocket and offered them to whomever he was talking to. At the same time, he introduced me to the main representatives of the villages.

After the introduction, Uncle Qiu rested awhile. Then he took out the Wang genealogy book of Baixiu Village with great care, and said that he had to clarify certain points about the Wang genealogy with an old man from another village. He sat down next to the old man, carefully turning the pages and consulting him. After a while, he came back to tell me that he had sorted out some problems about the exact generation of the branch to which the old man belongs after they split off from the Wang ancestry in Baixiu Village. Uncle Qiu has had only one year's education, but he is very clear about the relationship of the various families and branches of the Wang lineage recorded in the genealogy book. People of his generation pay a great deal of attention to the relationships among the different generations of the Wang lineage. They can tell who is the descendant of which family and who the descendant of which branch.

After lunch, the villagers followed the lion-dancing procession to the ancestral hall of Bailang Village to offer a sacrifice to the ancestors. The inside of the hall was empty because it had been ransacked during the Cultural Revolution. What was left were some faint antifeudalism slogans in red paint. The hall compared less favorably with the Grand Ancestral Hall in Baixiu Village; yet in the eyes of Uncle Qiu, the revival of lineage activities in Baixiu Village fell short of that in Bailang Village: "Although our ancestral hall has been rebuilt, our people do not respect it that much. It is often used for holding wedding banquets. This New Year, someone even suggested holding activities in the Fraternity Association; they seemed to have forgotten the ceremony of offering sacrifice to our ancestors in the ancestral hall. They had forgotten their 'first ancestor' [as worshipped in the Grand Ancestral Hall]. They are not united, and care only about their interests." It is always like this: when he talks about the lineage activities in Baixiu, Uncle Qiu seems bitter.

In fact, Uncle Qiu had told me of quite a few problems about the Wang lineage. The incident that annoyed him most was the moving of an ancestor's tomb. A Wang ancestor of the eighth generation was buried in the east part of Baixiu Village, beside the industrial area. If the tomb were removed, a large piece of vacant land would be left behind. Six or seven years ago, the secretary of the village government had suggested moving the tomb to another place. Uncle Qiu and some other old people thought that the suggestion was made to allow factories to be built in the area. At the time, most of the Wangs were against the proposal. However, by the end of 1994, some Wangs active in lineage affairs claimed that the tomb was bad *feng shui* and suggested moving it somewhere else. Uncle Qiu was flabbergasted.

"Why move the tomb? They have been bribed by the secretary," he sputtered. Then, composing himself, he changed his words, "Maybe they weren't that bad. But they are definitely wrong. What bad *feng shui*? The tomb's been there several hundred years, why must it be moved? They really wanted to develop the land, but they are ungrateful to their ancestor by doing that. Today, people will do anything for gain. They do not even care about the five virtues—'benevolence, righteousness, civility, wisdom, faithfulness'!" He grew agitated as he spoke. He did not say in so many words that these relatives were a front for the secretary when they sold the land, but his attitude was clear.

After the ceremony, the people walked around the village, led by the lion dance. As they walked, firecrackers exploded, adding to the festive atmosphere. We passed many ruined temples and tablets of the gods. The villagers explained the significance of each temple and tablet, as there was a story behind every one of them. They told of how these spirits protected and blessed their ancestors. At every location, the villagers would burn joss sticks and pray; some even brought offerings. They prayed devoutly for blessings to be bestowed on them.

After circling the village in a procession, we had dinner. Then it was time for Cantonese opera. The old men from Baixiu Village were most absorbed in watching the performance. People of the younger generation were not as attracted to it. They felt it was dull, and so in groups of three to five, they left for the karaoke nightclub nearby. It was not until ten o'clock that the performance ended. Dragging their tired bodies, the villagers from Baixiu Village boarded the bus to go home.

From this meeting, I got the impression that Bailang Village had developed in a way rather similar to Baixiu Village. The majority of the farmland had been filled to be sold or rented to merchants from the outside to build factories on. Not only in these two villages, but in villages throughout the Pearl River Delta, the pattern was the same. Any village located near a main road easily accessible to transportation would most likely be taken over for the construction of factories. This change has created many job and business opportunities for the villagers and has improved their living standards. In general, villagers, especially those of the younger generation, welcomed the new economic and open-door policies.

Faced with the drastic changes brought about by the reform and opening-up policy, Uncle Qiu feels that relationships among people can no longer be gauged by the closeness of blood relations. And other matters like marriage are no longer to be determined by parents. Young people not only do not listen to their parents, they might oppose and criticize them. Although lineage activities have been revived, some people make use of them for personal profit. All these new ways of doing things go against the traditional ethics that Uncle Qiu believes in. Obviously, Uncle Qiu has not been able to adapt to these changes in a short time, so he harbors a rather negative attitude toward reform and opening-up.

Uncle Qiu's Hopes

One afternoon, Uncle Qiu sought my help in moving some unwieldy furniture to his home. On the way, with the furniture between us, he said, "Last year, I should have taken you to my home for a dinner. But my home is not like a home. How could I invite you to come in, let alone have dinner there?" At first, I interpreted that to mean that his home was a mess since he lived alone. Later I found out that he did not mean that. "My home is not like a home" has totally different implications.

Uncle Qiu lived in the southeast side of Baixiu Village, in the region of old houses. His house is said to be over one hundred years old. It is quite a big house, with a courtyard of about 20 square meters. There is a big sitting room and three other rooms. Other than Uncle Qiu, another family, relatives of his wife's, lived there. Uncle Qiu occupied a room to the left of the entrance, over 10 square meters in size. It had no windows, and even in daytime one had to switch on the lights or the room would be very dark. The room was small in size, but was far from being messy; everything in it was quite neatly placed. In the room were a bed, a wardrobe—both very old-fashioned—a table beside the bed, and an ashtray full of cigarette butts on the table. There were also a few picture frames holding pictures of his son and granddaughters, and one of him and his wife when they were young.

I took the opportunity to ask Uncle Qiu why he often said that his home was not like a home. He answered, "I do not mean that my home is untidy. I mean that it is defective. My wife and my son are not with me, and I live alone. Is this a home? My life is bitter. What can I do? Well, it's been like that for years and years."

After putting the furniture in place, Uncle Qiu offered me a cup of tea. As I sipped the tea, I looked at the photos on the table. Uncle Qiu pointed to the photo of him and his wife; he said that as a young woman, his wife was very beautiful. Then he took out more photos from a drawer and showed them to me. Most of them were pictures of his son. More pictures, this time taken from a tattered but neatly kept plastic bag, were of him and Qiu Sao. The photos were fading but they were neatly kept. As he showed them to me, he gave a running account of when and where the pictures were taken, and memories of their life surrounding them.

Uncle Qiu's biggest wish is that his wife will come back to him. He often figured in his mind how the house could be rearranged so that his wife could live more comfortably. He often urged his son to persuade his mother to come back.

I knew that Uncle Qiu had made telephone calls to Hong Kong to talk to Qiu Sao, but she just ignored him. At the beginning of 1995, he said that he would try to contact her. Then an opportunity arose: It was his daughter-in-law's birthday, so he had a pretext to call. If he did not succeed in persuading his wife, he could try again in October when his son had his birthday.

Three months after, I asked Uncle Qiu whether he had made the call. He had a vacant look when he said, "Calling or not calling makes no difference." He fell silent. Then after a while, he said, "I did make the call last month. When she answered, she asked me was there anything important, and before I could finish what I said, she told me there's nothing urgent, and she hung up." I asked him if he would try again, and he said in a defeated tone, "No, I won't. She doesn't want to talk to me. Let it be! It's been so many years now!" He did not look at me as he said these words, but only stared at the marketplace outside the barber shop.

We can see from Uncle Qiu's participation in lineage affairs that he is someone who hangs on to the traditional familial ethical rules. His marriage was arranged by his father, and he grumbled about it; his marriage turned out to be unhappy, and he sighed over it. In spite of all that, he still thinks that a marriage arranged by the parents is more reliable than the free marriage that young people go after now. On the other hand, his temperament was stubborn and he insisted on seeing the marital relationship as being one of "men superior, women inferior." He therefore had a marriage in which husband and wife were more separate than together. To a modern mind, this marriage, which lacks the conditions for developing any love between the partners and is so encumbered by various negative circumstances, should have ended a long time ago. However, Uncle Qiu still fervently hopes that his wife will come back to live with him in Baixiu Village. This hope reflects to some extent that the continuity of traditional marriage is based not on feelings of affection between husband and wife but on traditional ethical rules.

5

The Story of Zhichao

The Transition from Tradition to Modernity

In 1949, after the Communists established rule over China, the government encouraged free marriage through legislation so that the traditional inequality between men and women could be eliminated. Despite that effort, in as late as the 1970s, the majority of the villagers in Baixiu Village found their partners through the intervention of a matchmaker or a relative. Some villagers in their forties still believe that a marriage arranged by the parents or a matchmaker is more reliable than a marriage in which partners choose by themselves. Neither the Marriage Law effected after 1949 nor political education seems to have had much influence over these villagers, who were born after 1949.

The protagonist of this chapter, Li Zhichao, was born after 1949. Compared to his peers, he was more forward in trying to select his spouse; however, this does not mean that he is not under the shadow of traditional familial concepts. Whether in the matter of marriage or in his family life, Li Zhichao is still deeply influenced by traditional familial culture. In this chapter, we will try to demonstrate how traditional familial culture continues its hold on the middle-aged generation. Through the story of Zhichao's marriage we will also see the long-term influence on this generation of the idea of autonomy engendered by the reform and opening-up policy.

Li Zhichao

Li Zhichao is twenty years younger than Uncle Qiu. He was married in 1979. His wife comes from Nanmen Village south of Baixiu Village. They have a son and a daughter. The son's name is Shijun; he was born in 1980. The daughter, Weiling, is two years younger. Zhichao's job is driving people by motorcycle to and from Baixiu Village. I got to know him quite by chance.

One day, after starting my field work in Baixiu Village, I had to go to Qingyang Town. The distance between the two places is about six kilometers. To take the small bus at the entrance to the village, I had to walk more than ten minutes. Instead, I called for a car. A motorcycle turned up, and Zhichao was the driver. He said the fare was 15 yuan. At that time, I knew little about the fares, but somehow sensed that it was on the expensive side. A few days later, I saw him again. He greeted me at once, saying "Hi Hongkonger, do you want to go to Qingyang Town?" There was a sly smile on his face.

I did not analyze that sly smile. Later on, I did take his motorcycle to Qingyang Town, and it was then that I understood the meaning of that expression on his face. For that trip, he charged me only 10 yuan. When I asked out of curiosity why it was cheaper this time, he told me straightforwardly, "The fare for a trip from the village to the town is normally 10 yuan." I was piqued. I asked, "Why did you charge me 15 yuan the last time?" He looked sheepish and said, laughingly, "I didn't know you then. As you are a Hongkonger, of course I charged you more." I soon found out that other drivers of motorcycles had the same practice of charging strangers higher fares. Zhichao was just a little more greedy than they.

The differential treatment accorded locals and outsiders is a very common practice indeed. When you go into a shop to buy something, shopkeepers will charge you more if they know you are an outsider. Sometimes, when I was commuting between Qingyang Town and Baixiu Village on the small bus, I saw motorcycle drivers bully the outside workers by asking for higher fares. In Baixiu Village and indeed along the coast of the Guangdong Province, this practice is prevalent. Besides wanting to make more money, what motivates this behavior is the clear distinction drawn between "us" and "them."

I gradually got to know Zhichao better. I often went to his house and spent the time discussing village happenings. Sometimes, I would teach his children English. In 1994, his son Shijun was in the second form in a secondary school. One day, when I was in Zhichao's house, Shijun was doing some homework on the subject of politics. One of the questions was, "As a good citizen, which do you consider more important: the interest of the state or the interest of the individual?" When Zhichao spotted it, he said to his son, "Of course the interest of the individual is more important, you moron!" Shijun, embarrassed, protested, "You can't give such an answer on the examination." Zhichao lectured his son, "I don't mean to tell you to answer like that;

I am simply teaching you how to be a wiser guy." Then he laughingly turned to me and said, "How can one make money following such nonsense!" He shook his head as he drew in a deep smoke. His expression was one of mockery.

Zhichao's emphasis on individual interest is to some extent the result of the "money first" attitude fostered by the reform and opening up. Once, Zhichao's brother and some friends pooled some money to start a factory. They invited Zhichao to be the general manager. In the course of work, his brother could not raise enough money on his own; he asked if Zhichao could help. Zhichao said yes, but after a whole month's canvassing, he got nowhere. Eventually his brother got together enough capital, but he never mentioned anything about getting Zhichao to be the general manager again. After the implementation of the reform and opening-up policy, money came first in the eyes of the villagers, and instrumental values were the guiding principle for most people. Where money is concerned, relationships between friends and brothers count for little, let alone the interest of the state.

After 1979, Baixiu Village was gradually transformed from a peaceful farming village to an industrial village full of investment opportunities. Almost all the villagers have given up farming. Some went to work in factories. Those who were more enterprising looked for small business to invest in, for example, running a general store or renting a fish pond for fish farming. Others with capital raised by relatives in Hong Kong started a factory or went into business in Qingyang Town. Still others connected to the rich and powerful in the town sought to get rich through various means. Under such circumstances, Zhichao is no different from other villagers in concentrating his thoughts on making money. When he has no customers, he sits under the big banyan tree, his eyebrows knitted, smoking a cigarette, lost in deep and anguished thoughts. More often than not, he is thinking of ways of investing and striking it rich.

Zhichao has little money, and he is not well connected with the rich and powerful in the village, so all he can do is to sit there scheming and dreaming, and enviously watching others make it big. Often he said, with an I-deserve-better attitude, "If I had money to invest, I would know of better ways and I would make a pile of money." Then he would give me a blow-by-blow account of how he would do better than so-and-so in this and that, and how he would have the guts to take risks. From what he said, it is clear that he had the makings of a gambler.

Gambling is rife in Baixiu Village. There are two secret casinos in the village. The customers range from those in their teens to those over sixty. Zhichao loves gambling. Although he is not rich (in better months, he makes about 3,000 yuan), he is a big gambler. He would put 50 to 100 yuan on a bet, sometimes even 200 to 300 yuan. One day I saw him lose 700 yuan within minutes. He told me that toward the end of the lunar year in 1993, he lost over 10,000 yuan in half a day. He had little left for the New Year.

Although Zhichao is a womanizer and boasts of having a "theory" on women, he takes good care of his wife. He drives her to work and drives her home after work. Once when Chao Sao (wife of Chao) fell ill, he was very worried. He took her to see the doctor; in addition, he went to see a *feng shui* master. He said that if it was the *feng shui* of the house that made her ill, he would spare no money to get the *feng shui* right.

Zhichao chose his wife. He got married after a period of free courtship. Although he cared about his wife, their love is not like that of young couples in the village; it is more like the traditional relationship between husband and wife of Uncle Qiu's generation.

Zhichao's Courtship and Marriage

Before Zhichao met the girl who became his wife, he had courted another girl. That affair happened just after his graduation from secondary school in 1973, when he was eighteen. Baixiu Village was still working in the mode of collectivization. The whole village was divided into six production teams. Zhichao belonged to the third production team (henceforth referred to as the third team). He was a platoon leader of the local militia, a rank equivalent to that of supervisor. While on the job, he met a girl whom he fell in love with, and courtship followed.

When Zhichao started courting, the country's marriage law had been in existence for twenty years. It encourages young men and women to practice free love and to choose a marriage partner by themselves. Yet very few young people of Zhichao's generation would look for a partner by themselves. Zhichao told me that when he was working in the third team, his peers seldom had enough courage to approach the girls in the team. They marry through the intermediary of a matchmaker.

Some of Zhichao's friends of about the same age told me that when they reached the eligible age, the village was still organized into production teams. There were three shifts a day. Work in the fields began at eight o'clock in the morning and did not stop until six o'clock in the evening. There were meetings of the production team practically every day. They had neither the time nor the opportunity to meet girls in other villages. Even when they had the time, there was no one to introduce them. They explained that times were very different then. Men and women were clearly separate, and when they met someone of the opposite sex, they dared not start a conversation.

There was another factor at work. Most of the villagers in Baixiu Village are surnamed Wang. Although the marriage law expressly states that people with the same surname can enter into marriage, parents often forbade the marrying of a female Wang, let alone one from the same village. This greatly decreased the chances of a Wang getting married to someone in the village. Some parents were even against marrying anyone in the same village; they did not want local in-laws. The reason is that if the two families are in the same village, should anything embarrassing happen, it would be known right away.

Some parents might have had no objection to their sons dating a non-Wang in the same village. However, if a man in a production team so much as conversed a little longer than usual with a female teammate, then word would get around to the whole team. Since people were not as open-minded then as the young men and women now, they were terribly shy and would feel embarrassed. The men lacked the courage to take the initiative to get to know the girls in the village, much less to go courting. In the end, they had to resort to the assistance of a matchmaker.

Zhichao spoke to us one day in the research center about his first love. "When I got to know her, I was platoon leader in the third team, in charge of part of the management of the team. I liked her a great deal, I don't know what it was about her that attracted me." He was less than ingenuous, as he had once told me that the girl was quite pretty. "Maybe it's fate! I wanted to see her every day. We were often together after work. Sometimes we went to the town to see a movie, but not often. In 1974 transportation was not so easy, and walking there takes more than an hour. So we spent the time in the rice field or on the grass patch, 'measuring the grass and getting a moon tan' as we called it." Then he added, jokingly, "because we did not tan ourselves enough during daytime!"

I could tell that Zhichao had really liked that girl from the way he talked about her, but they did not get married in the end. "Come to think of it, I really can't say why it ended like that. There was a period of time in which I didn't visit her, and another period in which she didn't come to see me, and that's it!" Zhichao's eyebrows tightened. He scratched his head as he recollected, as if he too was puzzled why their affair should end like that.

It was curious that he should not know how his first love affair faded, so I quizzed on him why he did not visit her for a period of time. He remembered that once the team had taken on a job outside the village for six months. When the job was finished, he hurried to his girlfriend's house. There, he found a man whom his girlfriend had invited. He said to me, nonchalantly, "When she saw me, she asked me to sit down. I could see what the situation was, so I sat for a little while and then left. Since then, I never went to see her again.

"Maybe that's it. I hadn't seen her for six months. The first thing I did on return was to visit her. Seeing a guy, a stranger, in her house really upset me."

He continued, telling me how he felt then. "Her parents knew about our dating each other, but they still got a matchmaker to find a partner for their daughter, and she also went for such interviews with prospective husbands. She refused every one of them. I learned later that the man I met was such a candidate introduced by a matchmaker. They were having an interview that day." Zhichao was clearly unhappy that his girlfriend went along with the arrangement.

"I thought to myself, if the guy I saw was not her new boyfriend but just someone introduced by the matchmaker, she should have explained it to me clearly. If she really cared for me and put me first, she should have come to see me, but she didn't." He was still agitated as he spoke. "When things got that far, what could I do? Only thing is not to see her again. It is not up to me to go and ask her: Do you really like me?" His eyes flashed a look of contempt, "If that guy were her new boyfriend, then wouldn't I lose face?

"After a while, I learned that she did not fall for that guy. I met her once by chance and she asked me why I hadn't visited her after that day. By then, I had met my present wife. Soon she got to know a new boyfriend through a matchmaker and eventually married him and moved to the adjacent village." Zhichao was a little regretful that things had turned out that way, but he did not regret that he had not tried to sort out the truth at the time.

We can see how a love affair that could have been built on mutual love collapsed. There was a small misunderstanding that could have been cleared if both parties talked it through honestly face to face. Zhichao's actions were to a certain extent motivated by the idea of "men superior, women inferior." This led to the sense of self-righteousness that was quite uncalled for and destroyed the love built up between the two. We cannot say that this male psychology is only present in traditional Chinese men. Modern-day men and women, be they Eastern or Western, also break up due to misunderstandings caused by minor events, especially those in which a third party is involved. Nowadays, however, a two-year relationship such as the one between Zhichao and his girlfriend would not so easily be broken. There would be confrontation and explanation, one or more times, rather than a silent dissolution based on one person's decision without a chance for the other to explain. This might demonstrate the firm hold that the familial maxim of "men dominant, women subservient" has on people's minds.

Several months after Zhichao and his girlfriend separated, he met his present wife Chen Xiaozhen (Chao Sao). Xiaozhen lived in a village south of Baixiu Village. Zhichao met Xiaozhen's brother at his cousin's wedding, and they became friends. The brother had just had a new house built. When Zhichao went to visit him, he stayed on to move the furniture and to clean the house, which was quite a mess. It was getting late and the road was dark. There were no street lamps in 1977. Xiaozhen then lent him a flashlight. Zhichao remembered clearly what he said to Xiaozhen as she handed over the flashlight, "I don't know when I can return the flashlight to you," and Xiaozhen replied, "You can do that when you come to visit us again."

Zhichao put on a sly smile when he reached this point of the narration. He said to me, "Any fool could see what she meant. She wanted to see me again." After that, Zhichao often went to Xiaozhen's home. He stayed late every time, so that by the time he left, it was dark, and the returned torch would have to be borrowed again. After a few such exchanges, they began a more serious courtship. Zhichao took the lending of the torch as a hint that Xiaozhen gave him to come back again. Zhichao found Xiaozhen very likeable; he liked her personable disposition and the friendly way she treated others.

The courtship lasted two years before they finally got married. During that period of time, whenever Zhichao had to leave the village to work, he would tell Xiaozhen he was going but would not tell her when he would come back. The reason he gave for doing this was that Xiaozhen lived far away, unlike his former girlfriend, whom he could see every day in the third team. The day he came back might not be the day they met again anyway.

"When I was platoon leader in the third team, I could arrange for her [the first girlfriend] to be in my group, so I could know what she was doing every day. Now Xiaozhen lived so far off, we saw each other only occasionally. I had no way of keeping her 'under surveillance,' so I would suddenly appear in her house to find out what she was doing." "Surveillance" was Zhichao's exact word. He continued, "Sometimes I would set a trap and say, 'I saw you with a strange man a few days ago,' just to see how she reacted. If she reacted in an agitated way, then it shows she did not have other boyfriends."

Zhichao would also put Xiaozhen's character to the test. "Sometimes I would ask her to go out for a movie on a certain evening, and then when the time came, I would tell her I was busy and could not go, just to see how she reacted. Usually she would say, 'If you're busy, then let us go some other day. Business comes first, and in any case, I have got work to do.'" Zhichao's view is that if she lost her temper because of the cancellation of a night out, then she would be a person without patience.

Although Zhichao selected his wife through romantic love, we can see that, from the way he described his courtship, he demanded compliance and patience from the opposite sex. In his first love affair, he expected his girlfriend to come to him to explain why there was a man in the house; he himself would not ask her because he felt he would "lose face" by doing so. He tested Xiaozhen in different ways in order to see if she was indeed an obedient woman. This way of thinking reveals his belief that men should be the dominant partner in the relationship between the sexes.

Zhichao's parents soon got wind of his dating a girl in the next village. Since they had not seen her before, they asked their relatives to find out if Zhichao was indeed going out with a girl and to see what sort of girl she was. This detective work was common among villagers of the older generation when they set out to select a wife for their sons. They were keen to find out the family background, the character, and the health of a would-be daughter-in-law. Word soon got back to Zhichao's parents that their son was indeed in love. When they learned that Xiaozhen's parents had both died, that she was living with her brothers, and that she was slim in build and weak in constitution, they were not happy. They thought she was no match for their son. Zhichao's mother, in particular, managed many times to dissuade Zhichao from marrying Xiaozhen. She sent for the matchmaker to come up with other girls as matches. What the parents and relatives were against was the fact that Xiaozhen was constitutionally weak. They worried about whether she would bear strong babies, thus continuing the Li lineage, and whether she would be a good worker, one who was strong enough to work in the fields as well as look after the house.

Even Zhichao's friends thought much the same way. They were of the opinion that if Zhichao married Xiaozhen, he would be in for a hard time because he would have to do the work of two people. Zhichao also gave the matter some thought. He realized that Xiaozhen was weak in constitution. He knew too that although she grew up in a farming village, she did not have any experience in farm work.

"They all told me to keep a clear head. Marrying her would mean hard work for myself. But what could I do? I liked her. When one is going steady with someone, one does not worry about such problems. Let the future take its course," Zhichao said calmly as he recalled what happened. But whenever he came to the part where his parents and friends were unhappy about Xiaozhen's slender build, he would add self-justifyingly, "I too thought her somewhat delicate, but she turned out to be very capable. She could do twice what a person could do in the rice field. How much do you think she could do? More than six *mu!* My friends in the village all said to me, 'We had no idea your wife was so fantastic.'" Zhichao said this with great pride in having made the right choice.

In the face of his parents' opposition, Zhichao married Xiaozhen in 1979. That was the year when smuggling people to Hong Kong was at a peak. Zhichao had tried to slip into Hong Kong before, but was unsuccessful. This time, he had the money but he was getting married.

"On the eve of the marriage, I had second thoughts. I felt a sense of reluctance; I did not want to marry her. Maybe it was then I found out that I did not love her as much as I thought. In actual fact, I still don't know why I married her. Maybe I pitied her, seeing that her parents had passed away when she was a child and she only had her brother to look after her. I had been going out with her for two years. If I left her then, she would be so much more pitiable. I felt I owed it to her. In the end, I decided not to try to migrate illegally to Hong Kong. Instead, I gave the money to my brother for him to have a go at it. Then I got married." Zhichao could not help breaking into a chuckle at this point. He asked me if I thought him silly, as if he did not quite understand why he decided the way he did.

In Zhichao's case, we can see clearly one big difference in value orientation and lifestyle between Chinese and Westerners. In an important life decision like marriage, Westerners base their decision on strong personal feelings such as love or hate, but the Chinese decide on the basis of social constraints and functions. This is a demonstration of the theoretical framework expounded in chapter 1. In the relationship between people, the Chinese emphasize continuity. For the Westerners, once there is no love, then there is no marriage. However, the feeling of love between two people will fluctuate over time. If questions arising from the love relationship are dealt with on the basis of the presence or absence of love felt, then the relationship will also fluctuate severely. A relationship based on social constraint and righteousness, on the other hand, will transcend individual likes and dislikes, and therefore the relationship will continue. Here, we are not assessing the pros and cons of the two lifestyles from a particular value orientation; we are simply pointing out that in our daily life, different concepts of the individual and social relationships will constrain and direct people's behavior and form different lifestyles.

The Years After the Marriage

Zhichao and Xiaozhen were married in 1979, the year China started the economic reform and opening-up policy. Like other villages, Baixiu Village adopted the policy of dividing the farmland and distributing it to each household. Each person in a household got two to three *mu* of farmland, and with the family becoming the accountable unit, the privatization of farming was quietly initiated. Zhichao, having a family of two, obtained a little more than six *mu* of land, to be used for farming purposes. From 1980 on, the system of production teams was all but abolished. Villagers could find work outside the team they belonged to. Zhichao was not to be outdone; he went to Qingyang Town to become a worker who moved goods. The pay then was about 7 yuan a day, and he stayed at the job until 1986.

Other villagers who had land distributed to them stayed to become farmers. Only people in those households with spare labor could try their luck outside the village. Zhichao now had a household independent of his parents, with only him and his wife. To get more income, he asked his wife to take care of all the six *mu* of farmland.

By 1986, more and more factories were built in Baixiu Village, so Zhichao came back and found a similar job in a factory as a mover. His monthly wages were about 400 yuan. In the morning and after work in the evening, he would go to the field to help Chao Sao (Chen Xiaochen). At night, he might follow the company bus to go to Shenzhen to do some extra work. From then on, the family's livelihood gradually improved. In 1990, their farmland was requisitioned by the government for industrial use, and Chao Sao went to work in a factory in Baixiu Village.

In 1990, Zhichao gathered enough money and, with some friends, went into the business of buying and selling salted fish in Qingyang Town. They made an average profit of 2,000 yuan a month. When business was good, they had a monthly income of 3,000 to 4,000 yuan. After one or two years, business slowed down because there were more and more companies doing the same trade. In addition, there were more and more bandits on the road. Going to Zhuhai with a lot of cash to buy salted fish became a very dangerous undertaking. In 1993, Zhichao decided to close his business and went back to Baixiu Village to become a motorcycle driver, transporting both goods and passengers.

Thinking back about the days after marriage, Zhichao admitted that his wife had been a great help at home. Chao Sao could do the work of two farmers when in the field, so he could leave her alone and went for another job outside the village. He said that not only were other women no comparison with his wife, even men would find it too tough farming six *mu* of rice field alone.

"One year after the wedding, she was pregnant. She still went to the field to work! Do you know how hard it is to farm? You've got to weed, to plough. I had no idea how she could manage it." At this point, the normally sly rolling eyes of Zhichao stopped flickering. He looked straight at me, and continued, "She never uttered any complaints! Not a single one!" She would come back to the house after working in the field, and cook and do housework. He felt that his wife was more capable than others; she could tackle any job. More importantly, she did not complain.

In 1981, Chao Sao gave birth to their first child. Because now there were three in the family, the household was given nine *mu* of farmland. Chao Sao worked alone still, and Zhichao continued to go to the town to work. "After the birth of Little Jun, she had to look after him, and at the same time cultivate the fields. She did it all by herself." Gone was his cunning look. Instead, his voice took on a deeper tone and his words bespoke gratitude for his wife's labor.

One can see from this couple the relationship between husband and wife in a traditional family. Zhichao said that he did not seem to feel love for Chao Sao, yet in his telling of their story, he expressed a sense of gratitude toward her. This gratitude was earned by Chao Sao for having discharged so well the duties expected of her from traditional ethics. From this, we see evidence of the traditional marriage concepts explained earlier, that marriage is a functional union. In the traditional concept of marriage, in the context of the demanding labor of a farming village, a husband, a mother-in-law, a father-in-law, and a daughter-in-law like Chao Sao each has his or her place in a marriage defined in relation to the ability to share back-breaking farm work, child care, and household chores. Other people too will judge the worth of a wife, even the success of a marriage, according to the same criteria.

In 1983, two years after the first son was born, Chao Sao gave birth to a girl. As Zhichao was still working outside the village, Chao Sao had to till the soil in the daytime and look after the children and the household work in the evening.

Zhichao said that at the time of the marriage, his parents did not like Chao Sao much. They gradually changed their attitude when they saw that the daughter-in-law was a good worker, deferential to the elders, and friendly toward others. Moreover, she brought a grandson to the Li family. The parents were overjoyed.

One day Zhichao invited me to this home to have fish congee. That was the first time I met Chao Sao. She looked quite different from what Zhichao made her out to be. He had said she was small in build, and that was why his parents did not like her. Chao Sao was indeed a little shorter than Zhichao, but she was healthy, with a pair of strong arms. Zhichao later explained that working in the fields had made her grow tougher. Chao Sao was dressed simply. She had on a pair of grey trousers and a light blue blouse. I noticed later that the outfit was not only for going to the factory to work; it was the same as she wore at home. Her face was rounder than Zhichao's, so the eyes were comparatively smaller. Her hair reached just to the shoulders, and was tied up with a rubber band. My first impression was that she was somewhat older than Zhichao.

Chao Sao greeted me courteously, then said, "Teacher, we have got some congee ready. I wonder if you'd like some." I exchanged some courtesies and sat down to have a chat with Zhichao. Chao Sao did not sit down with us but went back to the kitchen to prepare more fish congee. Soon the congee was ready, and when we were eating, she did not say much, only to ask if the congee was to my liking. I said it was indeed, and she appeared very pleased. Then she said, "If that's so, come more often to have congee." Then Zhichao revealed that she had bought two catties of fish and that was why the congee was so tasty. When the meal was over, she cleared the table and went back to the kitchen to work; she did not stay behind to chat.

At first, I thought that her taciturnity was due to my being a stranger and that it was my first visit to the house, but after a period of time, when I got to know the family quite well, she was still always busying herself with household work. It could be that she thinks entertaining friends is the job of the husband and that her station is that of a housewife.

When Zhichao's farmland was requisitioned to make room for factories, he could not support the family just by his work. Chao Sao did not work in the paddy fields any more. Rather than staying idle at home, she went to work in a factory, just like other women from poor families. She worked overtime several times a week. After a full day's work, she would go home and do all the house chores. She must feel that she has done all that is required of a wife and daughter-in-law.

Chao Sao, the "Gentle and Virtuous"

When I first met Chao Sao, her figure and her dress immediately brought to mind what Zhichao said to me one day. He was talking about the criteria he used for selecting a wife. "I hate women with a burly figure! No matter how pretty she looks, I really do not go for women who are stronger in body build than I am!" He frowned as he smoked. "My parents and their friends got it all wrong. They were keen to introduce women who are big in size compared to me. I just dislike women like that."

Before 1979, if you had a wife with a well-built body, she could do much work in the fields. Of course peasants liked to have such help. That was one of the criteria for the selection of a wife or a daughter-in-law. Zhichao did not like girls of that type, and chose instead the slim Chao Sao. Ironically, after marriage, because Chao Sao had to till the soil, she became strong and hardy.

Chao Sao seldom used cosmetics or wore stylish clothes. Whenever I saw her, she was in the same factory uniform. Sometimes the uniform was soiled and she seemed not to have time to wash it. Even on holidays, she was dressed in simple, plain clothes.

I think that Chao Sao's figure and her appearance are a direct product of her role in the family. The image she projects coincides with that of women in the Baixiu Village in the pre-reform days. Chao Sao and women older than she would think that to be a "virtuous" wife, appearance is of little consequence.

Faced with a wife who is dressed plainly, who is rather robust, and who works hard without complaint, Zhichao has mixed emotions. He does not like burly women. From a traditional point of view, if Baixiu Village had not modernized under the reform and opening-up policy, then without doubt, Chao Sao is an ideal woman who has satisfied the traditional normative requirement of being "virtuous." Her physiological change, from a slender woman to a robust one, is a mark of her devotion to her family.

But Baixiu Village in the 1990s is no longer a peaceful village with a simple way of life. Economic changes over a decade have transformed the village. The young women in the village are no longer content with dark-colored blouses and khaki trousers. They have begun to follow the fashionable trends as seen on television or other sources. In Qingyang Town, shops and a variety of goods abound. Fashionable women's clothing and all sorts of cosmetics are readily available. Not only are young women concerned with their appearance, even housewives in more well-heeled families dress in fashionable clothes. Plain clothing is no longer the norm for women in the village, and no longer the mark of a "virtuous" woman. Chao Sao, who is dressed in much the same way as women of older generations, now looks even more unsophisticated, behind the times.

A new factor has come into the picture. After a large number of factories were built in Baixiu Village, a large number of female workers came from outside the village. Some of them were attractive with voluptuous figures. They became karaoke nightclub girls. They were fashionably dressed, wore heavy make-up, and exuded feminine charm. The young male villagers might have girlfriends or wives who were dressed somewhat stylishly, but there was no comparison to these newcomers. Few male villagers could resist these women. Zhichao, about forty years old, was no exception. When he met a beautiful woman worker, or a heavily made-up karaoke nightclub girl, he would eye her lasciviously.

Chao Sao's appearance is outdated, and compared to these women, she is matronly. Zhichao has never said anything to indicate he is ashamed of his wife's appearance. Quite the contrary, he is explicit in his pride in having a good wife. But he is of the belief that he does not really love her. He desires, like richer people in the village, to "have" a woman or in the slang of the district, "buy out a concubine" (pay for her living expenses). Under the circumstances, finding a woman whose looks are better than Chao

Sao's to satisfy his lust, is an easy matter. If Zhichao has money, then he can easily find, among the sing-along girls in karaoke nightclubs and women workers from other provinces, petite women that he fancies who could become his "concubine." If this happens, Zhichao's feeling that he does not love his wife will be reinforced.

When Chao Sao was not too busy, I tried to strike up a conversation with her so as to understand more deeply the life of women in Baixiu Village. Most of the time, though, she answered my questions tersely. At first, I thought she was introspective in character, someone taciturn by nature. Later, by chance, I found she was not introspective, as I had thought. She once invited fellow workers from the factory home for dinner. She was quite forward in entertaining them. She chatted cheerily and humorously, more so than when she was with Zhichao. Later still, I found that she quite liked playing the hostess, and often invited women workers home for a meal.

In a village, husband and wife often have separate circles of friends. The man will mix in a group of male relatives and friends; the woman will have her own group of female friends. In Baixiu Village, a group of men are often found talking in the marketplace, or chatting in a circle just outside a shop, or huddled around an old American pool table playing billiards or gambling. In front of grocery shops, there will be a group of women chatting away.

The old people, for example Zhichao's father, rise early and go to the tea house in the next village for tea. There, you will find a crowd of old people all about the same age, all male. Generally speaking, it seems that village men of older generations spend the whole day chatting with each other, and do not go home except for meals.

Zhichao's marital relationship does not differ much from those of his father's generation. Even when there is no business, he will not go home, but instead sit down under the big banyan tree in the marketplace with a group of motorcycle drivers, engaged in long conversations. Or he might gamble with other drivers or villagers. The relationship among friends and colleagues in the same-sex group is more intimate than that between husband and wife. At home, Zhichao and Chao Sao each has household chores. When I was in Zhichao's home, even when both of them were watching television or taking a rest on the sofa, they seldom spoke to each other. All I could hear was the sound from the television or the whirring of the electric fan on the ceiling.

On the other hand, it would be wrong to say that Zhichao and Chao Sao are devoid of feelings for each other. By comparison with male villagers of his generation, Zhichao shows concern toward his wife. The feelings they have for each other, as explained earlier, are those of traditional couples:

each fulfills his or her expected duties. They have gone through a period of hardship, and this experience constitutes for each an obligation toward the other. The modern couple's "romantic" feelings for each other appear to have surfaced before the marriage, but they have not been evident after it.

Although Chao Sao does not have to go to the paddy fields to work, she is very busy most of the time. Every morning, she gets up early, goes to the market to buy food for lunch and dinner. After going home, she makes breakfast for Zhichao and the children. Then she goes to the factory to work. At noon, she goes home to cook lunch and in one hour, finish cooking, eating, and cleaning; then she returns to the factory. Four days a week she works overtime and does not get home until nine o'clock. It is only then that she can do the housework.

The working hours of Zhichao's transport business are much more flexible, but he has never helped Chao Sao in her housework. Even when business is slow, he will sit in the usual spot in the marketplace in the village and prattle on. Though he knows that Chao Sao is extremely busy cooking and doing housework when she gets home, he does not go home earlier to help her prepare a meal. Sometimes, when he gets home late and eats what Chao Sao has put aside for him, he will put the plates and bowls into the sink after eating, leaving them there for Chao Sao to wash when she comes home after work at night. When he is home in the evening, he will plant himself on the sofa and watch television. When talking about how hard Chao Sao works, Zhichao boasts about her capability. Then he would talk about the old days and how much she could do in the paddy field by herself and look after the family at the same time. Now she is enjoying an easier life!

Measured against a modern society's standards, Zhichao's share of the housework shows him to be far less concerned about his wife than modern husbands are. However, judged by traditional standards, Zhichao can be considered a family man. Traditionally, the Chinese do not care much about personal feelings. They look at human relationships more from the viewpoint of societal rules and functions. For example, Zhichao realized before marriage that he did not love his future wife, yet he suppressed his personal feelings and fulfilled the promise of marrying Chao Sao. After marriage, he did not pay much attention to his wife's feelings but was punctilious in meeting the functional needs of a family.

Zhichao was born in the 1950s and grew up under Communist rule. In him we see few traces of the effects of the marriage laws of the 1950s; likewise we do not see evidence of political movements and education in the way he dealt with the issue of marriage and the way he deals with family affairs. He is not much different from people of an older generation. His marriage is self-made, but in making the decisions involved, he did not base them on personal feelings. To a large extent, his courtship and marriage show that he is a believer that the relationship between husband and wife is one of "men dominant, women subservient." The story of Zhichao demonstrates that the new marriage system espoused by the Communist Party did not change the traditional ideas about marriage of the middle-aged villagers of Baixiu Village. On the contrary, traditional familial culture is still deeply rooted in the minds of people of that generation.

6

The Story of Wang Qiming

A Marriage in the 1990s

In the 1990s, hotels, restaurants, and all types of shops sprang up like mushrooms in Qingyang Town as the economy prospered. They provided locals and workers from other provinces with many job opportunities. In Baixiu Village, the number of factories increased to about forty, and nearby, many shops selling various merchandise were opened. Only one-third of the farmland remained and some of it was even uncultivated. The shift in economic patterns has provided more opportunities for men and women to get together both at work and in social settings. The separation between men and women has become less strict. People are no longer shy in talking to strangers of the opposite sex. Compared to the days of Zhichao, it is much easier for young people to find a partner. There has been an enormous change in the ideas about courting and marriage in the sense that young people can date or marry someone of their choice. However, there are some old people, like Wang Zhenqiu, who still follow traditional values when looking for a wife for his son. Hence, conflicts between the generations arise on the question of marriage, because they have different standards. Couples married for romantic love have a more intimate relationship. Husbands are more attentive to the personal feelings of their wives. To a certain extent, this changes the family structure in that the core of a family shifts to husband and wife. Such a change lessens the impact of a traditional patriarchal culture and could possibly close the gap of inequality between the sexes. On the other hand, the spread of the sex trade in the 1990s coupled with the male villagers' indulgence in sex has caused further changes in the relationship of the sexes. In this chapter, through the description of the courting experience and marriage of Wang Qiming, a young villager, we hope to illustrate how the tensions created by the development of romantic love, the impact of the traditional patriarchal culture, and the rampant growth of the sex trade affect the young villagers on the issues of love, marriage, and the conjugal relationship.

Wang Qiming: Coming of Age After the Reform and Opening Up

Wang Qiming was born in 1969. After finishing senior middle school, he worked in a factory earning 400 yuan a month. This was considered a fairly good salary at that time. But he thought it was a job without many prospects. Most of his friends in the village were able to seize the opportunities opened up by the rural economic reform. Some ran small businesses in the village while others went to Qingyang Town to do business. Qiming worked in the factory for less than a year. At the end of 1992, he ventured out to Qingyang Town to open a shoe shop. Business was good at first. He made a profit of 3,000 to 4,000 yuan a month. In 1993, economic development was at its peak. While more and more shops opened, people started to make money by reselling shops at unreasonably high prices; scalping was the craze. Qiming found that some people made tens of thousands in just a couple of months. So, he got a loan from his father and jumped on the bandwagon. The shoe business was no longer important to him. In this year, more and more people turned their farmland into factories and sold them to foreign investors. As a result, the business of excavation and transporting soil to level off the farmland flourished, thus driving up the prices of trucks used for transporting soil. Qiming and his cousins bought a truck for more than 200,000 yuan in the hope of making a profit in the resale. Unfortunately, after 1994, the fervor of economic development ebbed. The prices of shops went down. Qiming got rid of his three shops at a loss. At the same time, the business of transporting soil was on the decline. As a result, he lost all the money borrowed from his father. Looking back, Qiming admitted that he was too young then, just twenty-four. He could not resist the temptation of making big money. His inexperience in doing business and the eagerness for quick success contributed to his downfall. In the end, he closed his shoe shop and went back to the village to work in a factory.

Wang Meiling, Qiming's wife, grew up in his village. They were classmates in the primary and junior middle schools. She was poor, with seven siblings. Despite the fact that she was qualified to go on to senior middle school, she had to drop out and find a job to support her family. On the contrary, although Qiming did not do well academically and even stayed in the same grade for another year when he was in junior middle, his father, like many parents, offered bribes so that Qiming could continue his studies.

Having finished middle school, Qiming met Meiling at an alumni gathering. At that time, she was working as a supervisor in a factory in the village. Qiming's classmates and friends felt that Qiming and Meiling would make a good couple. They encouraged him to court her, so he took the initiative by participating in the alumni activities in the hope that he would see her and learn more about her. After attending several group functions, they started going out by themselves.

When Qiming recalled why he liked Meiling, he said, "She was the most beautiful girl in the alumni association. Ever since I first met her, I liked her. Looking back, although I knew her before, I did not know much about her." Meiling has delicate features and attractive eyes; she is a good looking girl.

When Qiming recalled his courtship, he looked proud and said, "I didn't attract her attention purposely, nor did I give her gifts to please her like others. I asked her out and she didn't turn me down. I knew that she was disposed to like me." Qiming is 170 centimeters tall, muscular, and quite handsome. When I first met him, he had a trendy haircut, and a pair of old-fashioned black-rimmed glasses resting on his nose. Later he changed them to a more up-to-date pair with a gold-plated frame. After work, he liked to put on a T-shirt and jeans, which was the fashionable outfit of the young villagers.

In 1992, when Qiming was dating Meiling, he went to a karaoke nightclub with his boss. He said, "At first we went with our boss because we could not afford to go by ourselves. The nightclub was air-conditioned and we could eat, sing, and have the company of the karaoke 'ladies.' That's cool. You know, I'm just a village boy. So I was really excited to be there." He smiled slyly and said, "You can even choose your own 'lady' there."

When he first went to a karaoke nightclub, Qiming was just an ordinary staff member earning 400 yuan a month. For a twenty-three-year-old middle school graduate growing up in a village and working in a factory, going to karaoke nightclub was a novel and tempting experience. Unlike his fellow villagers, he could not afford to go there a couple of times a week. Nevertheless, when his business failed, he was rather depressed and passed the time in karaoke nightclubs.

"I met a girl in a karaoke nightclub. At that time I was dating Meiling. Of course, I wouldn't let her know. I thought if I met a girl I liked, I didn't consider it a problem to go out with her." Actually, Qiming and Meiling were very steady then and were talking about marriage.

Conflict Between Two Generations

After courting for three years, Qiming and Meiling got married in 1994. As his business failed the year before their marriage, Qiming did not have enough money to build a new house as a matrimonial home. They had no choice but to live with Qiming's parents. Qiming knew very well that this would cause a lot of problems. His parents did not like Meiling in the first place. After the marriage, they always had quarrels with his parents. Qiming felt helpless.

In an interview, Qiming told me, "My wife is more independent and has her own opinions. She is not submissive, nor does she follow suit in doing things. She is not superstitious and does not worship the gods. Yet my mother expects Meiling to pay respect to the ancestors during festivals, but she is reluctant to do so. When we were first married, this kind of conflict happened all the time. She never compromised." Qiming thought that his wife had done nothing wrong. It was his mother who was too stubborn.

Qiming complained about his mother to me. "My mother not only expects Meiling to follow the tradition, she is also highly demanding. Meiling is required to do the housework early in the morning. If she gets up late, my mother will be very unhappy and shows it."

As a matter of fact, Mrs. Wang is not that demanding according to the standards of her generation. Her expectations are just some basic requirements. So, if we judge Meiling by traditional family standards, she is not a dutiful daughter-in-law. Perhaps that is why Mrs. Wang's attitude toward her was getting worse.

Qiming told me, "I like to take Meiling out. Sometimes we come back late. Mother won't bug me, but she'll take it out on Meiling. She'll say the dinner is cold and there's no point in keeping it and she has to throw it away. Now, unless it's really something important, we'd rather not go out in the evening. Even when we go out, we try not to come back very late so that Meiling doesn't have to put up with all this."

Obviously, Qiming was troubled because his mother had made things difficult for his wife. He cherishes his emotional relationship with his wife. It is difficult to find this attitude in a middle-aged couple such as Zhichao and his wife.

"I wish we could live our lives, just the two of us. It's been a few months since we last went away for pleasure together. We went to town to see a performance by a Hong Kong singer. It cost me more than 600 yuan for two tickets and the return taxi fare was a few hundred yuan. In all, we spent almost 1,000 yuan. It's expensive, but Meiling likes the singer. Later, I'd like to take her to the Windows on the World in Shenzhen. People say the miniature constructions are marvelous and look so real."

Mrs. Wang did not appear to be very happy about the fact that Qiming and Meiling were so affectionate. What's more, she proposed cooking separate meals one year after Qiming got married. At first, he was a bit reluctant, but he eventually gave in. He told me, "Of course, it's a little bit embarrassing, a family sharing the same kitchen but cooking meals separately. But I can't think of any better way. My mother is really demanding. Even when we try our best to accommodate her, it'll certainly affect my relationship with my wife in the long run. Anyway, it has been much easier for us since then. We are freer. If we go out with friends, it doesn't matter when we have dinner. We don't have to be bothered by what my mother thinks." Qiming felt happier and it seemed as if all his problems had been solved.

Problems of Romantic Love

Well before Qiming and Meiling started courting, Qiming's parents had objected to their going out. Even the elderly and male relatives of his own generation often stood in their way. Once, when we were in a cafe, he revealed his problems to me.

"Not long after I began dating Meiling, some peers who were my seniors came to me and said, 'Meiling is not a good girl. She dated a lot of men, even had a dubious relationship with her boss.'"

The allegation that Meiling had had this dubious relationship arose mainly because she was promoted. She initially worked as a cashier earning only 300 yuan. But two years later she was promoted to the position of supervisor and got a raise of 300 yuan. At that time, Qiming had just got a job in a factory and did not earn as much as Meiling. He was paid about 400 yuan a month. According to Qiming, after Meiling was promoted, some men who were her seniors became her subordinates. They resented it. They thought a woman could not be competent enough to become their supervisor, so this rumor was started. Some people—even the boss from Hong Kong—believed she must have had a dubious relationship with the director.

The elders passed this rumor on to Qiming's parents. They were very unhappy, particularly his mother. She thought Meiling was not a good girl and asked Qiming to stop seeing her.

Qiming kept shaking his head and said, "I trust Meiling's character. My parents' disapproval and the rumor put a lot of pressure on me. No matter what, I continued to go out with her. Because of this, I had numerous fights with my mom. My uncles and aunts also tried to talk me out of this relationship, not to mention the male relatives who were my senior. I was very upset. Actually, dating is something between the two of us, but people stand in our way."

In addition to being from the same village, Meiling and Qiming both bear the surname Wang. Generally speaking, seniors do not want their sons to marry girls from the same village, let alone those having the same surname.

"I had a serious row with my mother on the question of whether people from the same village and having the same surname can marry. My father is more understanding and only mentioned this once or twice. Since I was very persistent, he didn't mention it again. However, my mother just couldn't leave me alone and dressed me down from time to time. She drove me crazy. As a matter of fact, if we go back to the question of ancestry and blood relations, Meiling does not belong to our lineage despite her surname Wang. We have to trace back to more than ten generations before finding a common ancestor. Moreover, Meiling was adopted and her surname was Li originally. I explained this to my mother several times. Though she sometimes accepted my explanation, she slipped back into her old ways after a day or two."

In the past few years, more and more Wang young men and women got married in Baixiu Village. For the young people, it is no longer a problem to marry someone with the same surname so long as they do not have a close genealogical relation. According to the marriage law enacted in 1950, couples can marry when they do not have any direct blood relation for five generations. Some senior villagers told me Baixiu Village has a similar custom, which allows marriage between villagers who have the same surname but no ancestral relation within five generations. Nevertheless, most villagers still uphold the tradition of "no same-surname marriage." Before 1949, marriage was arranged by parents. When they looked for a spouse for their child, they had already eliminated those having the same surname.

It was not until 1960 that the first couple having the same surname married in Baixiu Village. The next year, another couple having the same surname got married. The groom was named Wang Zonglin and the bride Wang Guozhang. At that time, their parents disapproved of the marriage. The bride's father even vowed to kill both of them. Regardless of their parents' disapproval, the young couple registered at the commune. They did not go back home to hold any wedding ceremony and spent their honeymoon in Guangzhou City.

Twenty years later, Wang Zonglin's child is now married. Whenever his mother recalled how he disobeyed his parents, she would take him to task. Wang Zonglin felt that his mother had not forgiven him.

This kind of marriage seldom happened in the 1960s. In Baixiu Village, the enactment of the marriage law has not brought about any fundamental changes in the idea of traditional marriage. Yet in the 1980s and early 1990s, same-surname marriages in the village have increased. Generally speaking, parents still object to same-surname marriage. However, for young people like Qiming and Meiling, when it comes to the question of choosing a spouse, they no longer submit to the pressure.

Anxieties About Building a House

After Qiming and his wife began cooking their own meals, they had fewer clashes with Qiming's parents. Nevertheless, the fact that the two generations live in the same house does cause discomfort. Qiming longs to build his own house so that he and Meiling can really lead their own lives.

Since 1949, the government has encouraged newlyweds to build their own house to avoid living with the in-laws. This way, in-laws cannot treat the daughter-in-law like a servant. In Baixiu Village, generally speaking, a couple will get their home ready before they marry. For example, Li Zhichao had already found a piece of land on which to build his new house before he married. Around 1979, it did not cost much to build a new house because there was an ample supply of land. However, the main reason that Zhichao built a new house was that he wanted to get more land. Not only did his wife not have any clashes with his mother, they got along very well. Zhichao knew that farmland belongs to the state, but that residential land becomes private after the owner obtains the title deed.

However, after 1990, land for residential use became scarce and the price was rather high. It was not easy for newlyweds to own their own house. Given that his business failed the year before he married and he had financial problems, Qiming had to live with his parents after marriage.

Although Qiming's salary was raised to more than 3,000 yuan in 1995, he still could not afford to build his own house. It really upset him. "It's not easy to build a house. One has to have at least 200,000 to 300,000 yuan. Given my present salary, I don't know how long it will take to save up that amount of money. By that time prices will have gone up."

As a matter of fact, an ordinary two-story house only costs a little more than 100,000 yuan. Lately, the well-off Baixiu villagers have begun to decorate their houses luxuriously. In addition, they like to make comparisons. Under this influence, Qiming, being senior staff, felt that he had to build a presentable house. He can only do so when his financial condition permits. It seems it is not possible for the time being, so building a house becomes Qiming's mental burden. Perhaps it is because he wants to build his house as soon as possible that he gambles a lot behind his wife's back. Every time his wife finds out, they have a quarrel.

"A couple of nights ago, I told Meiling I was going to visit a friend. She had already guessed that I was going to gamble. I got home after eleven. She was very unhappy and had a fit. I didn't answer her because I had done wrong. So I let her yell at me. But she was getting more furious and started crying. She packed her clothes and shouted, saying she would leave me. I spent a long time talking her out of that. In the end she stayed, but she's still very mad. She asked me to rent her a place in town and let her live alone. She didn't want to live with me, a gambling addict."

Qiming was crestfallen and continued, "Actually this is not the first time that we have fought over my gambling. It's just that this time she's really mad. Two weeks ago, we joined a tour organized by the rural government. I played games with some friends on the coach. She was upset and got furious when she saw me lose 4,000 yuan. She asked me not to gamble ever again, otherwise she'd never forgive me."

Qiming was remorseful about his addiction to gambling and said, "Our first and foremost wish was to build a house after we got married. It's very important, especially for my wife. This will reduce the clashes between her and my mother. Originally we had more savings, but I lost them. Well, I really let her down, so she has every reason to be mad." Qiming kept blaming himself. According to Zhichao, Qiming gambled big but had little luck. He lost about 10,000 yuan in one morning.

Men Decide, Women Follow

One evening, I invited Qiming and Meiling to dinner at a restaurant in another village. As soon as we had ordered, Qiming took out a Hong Kong horse racing post from his pocket and told us the night race had begun. After flipping through the paper, he went to the telephone booth to call to place a bet.

Meiling shook her head without saying a word. When Qiming came back, he saw me looking at him with a puzzled look, seemed to understand, and winked at me. After dinner, Meiling went to the washroom. I took this opportunity to ask him why he openly went to place his bet in front of Meiling. He smiled, "She has to accept the fact that men gamble. Anyway, I won't place a heavy wager." Qiming's attitude and tone suggested to me that Meiling had to yield to him eventually.

Qiming's attitude toward his wife reminded me of something he once said: "When we were dating, Meiling was very accommodating and gave me a lot of freedom. At that time, I'd already known that she would be a very considerate wife in the future." He had the following opinion when he compared local girls to those coming from the other provinces: "Outside girls are very different from local girls. Local girls are very diligent, submissive, and considerate. When it comes to decision making, they will put their husbands first and consider their opinions first. But the outside girls are different. They only think of themselves. They are less hard working and like to seek pleasure. So, when it comes to picking a wife, I won't marry an outside girl." Although Qiming had a bias against female workers coming from other provinces, this obviously tells us what he asks of his wife.

Qiming's view generally represents that of the young people in the village. Ah Song, his kin, is one of them. Ah Song told me he was looking for a girl who had similar interests and was on the same wavelength. Unfortunately, he had not come across one yet. When he talked about how couples got along, he was very traditional: "If a wife does not do the housework, or obey her husband, what's the point of marrying her?"

From Qiming's praise for his wife, we have a better idea of what he asks of her. He said, "My wife has another virtue. She takes care of everything at home, big and small, and I don't have to worry a bit. Frankly, it's the wife's duty to take care of everything at home, otherwise she is not a good wife."

A few months after Meiling gave birth to a baby, she did not want to stay at home. She got a job in Qingyang Town. It paid well, about 2,000 yuan. She worked twelve hours a day and got off at nine in the evening. Qiming found the hours to be too long. At first, I thought he felt it was too hard on her. Subsequently, I found out that he was afraid that Meiling would not have time for the housework. Evidently, he has never thought of sharing the housework with her.

Women of the older generation who marry into wealthy families have to take care of everything at home despite the fact that they have servants to attend to them. In the morning, the daughter-in-law has to get the toiletries ready and attend to the mother-in-law by helping her wash and dress. If she gets up late, she will be blamed for being disrespectful. The servants will do the cooking, but the daughter-in-law has to help to prepare the meal. A woman who marries into a poor family has a much harder life. Apart from helping the mother-in-law with the housework and attending to her husband, she also has to work in the fields. After 1949, the government launched the women's movement and attempted to educate women to protest against "unequal" treatment. However, it did not change the traditional status of women. When the parents picked a wife for their son, they still based their choice on the criteria of capability, being considerate, and obedient. As for the son, he not only wished for a beautiful wife, but also generally agreed with his parents. When Zhichao picked his wife, he emphasized the importance of her being considerate and extremely obedient. At first, he did not pay much attention to his wife's ability. Yet, after talking to his friends, he regretted that his wife was petite and weak. Qiming belongs to the younger generation and was able to choose his wife on his terms. However, it seems he still adheres to the view of the older generation when it comes to married life—men decide, women follow.

"My wife wants me to make the final decision on all important matters. I think it's absolutely right. Making important decisions should not be the woman's job. Even if she is not satisfied with my decision, or not very happy about it, I'll still insist. A man who cannot make decisions is absolutely useless," Qiming affirmed.

The traditional patriarchal culture and the rise of the sex trade in the nineties have had an impact on Qiming. In an interview he said, "I've gone to the karaoke nightclubs and it's nothing unusual. My peers of the lineage do the same thing, but I'm not a frequent patron."

"Did you have sex with the ladies?" I asked.

He did not answer me directly, but said, "Since I got married, I have seldom gone there. Sometimes, it's for the sake of business. I have to accompany my clients or the people from the *danwei* to the karaoke nightclubs for pleasure or on business. Sometimes I have to go with the factory boss or the manager. It all depends, a few nights a week, or a few times a month."

"In that case, you are not a frequent customer?" I followed up.

"'Having fun with women' is not really a problem. The thing is your normal life should not be disturbed." Qiming started to give his opinion about "having fun with women." "It costs a lot of money to go to the bars. I'm not that well off. 'Having fun with women' is a matter of chance. I think it's better to come across somebody you like naturally."

Qiming knows very well that if Meiling knows of his dalliance, she will ask for a divorce. Yet he has not refrained from going to nightclubs; he only does it more discreetly behind her back. When he was dating Meiling, he went out with a karaoke girl.

Qiming and Zhichao are very similar in this regard. Zhichao will never let his wife know. Men of Zhichao's generation do not think that they have done anything wrong to their wives, nor will they worry about their wives' divorcing them—they know very well that their wives will not. In their opinion, "having fun with women," even "keeping a second wife," is insignificant as long as they give their wives enough money to pay the bills. If a wife starts squabbling with her husband over this, she is unreasonable. Qiming sometimes expresses his agreement with Zhichao.

At the end of the eighties, karaoke nightclubs and similar establishments emerged in Baixiu Village. Their existence lured the male villagers to indulge in sex. The pursuit of romantic love has grown in the young people of Qiming's generation. To a certain extent, they attach great importance to the relationship of husband and wife, more than their parents do. Hence, this develops into a family pattern that makes husband and wife the core of a family. This kind of pattern might to a certain degree change the "men decide, women follow" relationship between men and women in the patriarchal culture. Nevertheless, from Qiming's marriage, it seems that the younger generation still inherits the traditional male-dominated culture. In addition, the rampant growth of the sex trade has an impact on Qiming. Subconsciously, he treats women as a commodity, like the other male villagers. Despite the fact that the young men wish for a marriage based on romantic love, the traditional value of a patriarchal culture, coupled with the temptation of sexual desires, hinders the development of an equal relationship with their wives.

Conclusion: Change in the Concept of Marriage

We have portrayed three marriages of different generations in Baixiu Village. In this section, we shall briefly sum up how the traditional familial culture and autonomy changed the concept of marriage for the three generations.

The protagonist in chapter 4, Uncle Qiu, grew up under the regulation of traditional family ethics. Not only his marriage, but also his life-sustaining barbershop business, was arranged by his father. From his story, we can see that his marriage, which was arranged by "the dictates of the parents, the words of a matchmaker," was a painful experience.

Since the reform movement, people took a different view of marriage. Living in the 1990s, Uncle Qiu notices that young people can now freely marry a partner of their choice. Occasionally, this stirred his emotions. He would blame his father for the arranged marriage. However, when he talked about the difference between freedom to marry and the traditional marriage, he often criticized free love as infatuation. The young people's choice was one-sided since they had not taken into account the well-being of the family. Uncle Qiu's criticism reflects how he is affected by his life experience in the 1990s. It has more or less undermined his belief in traditional values over the years. However, under the constraints of familial norms, the personal feelings can hardly persuade him to accept the ideas of romantic love and freedom to marry. In the 1990s, Uncle Qiu still upholds the value of traditional marriage.

The protagonist in chapter 5, Zhichao, who was born in the 1950s, grew up under the influence of the "antifeudal" and antitradition political movement. He chose his own wife. To a certain extent, this put into practice the idea of freedom in marriage instilled in the people by the government. Compared with Uncle Qiu, Zhichao has greater independence. However, judging from his dating process and married life, his choice was to a large extent affected by traditional values.

Zhichao had a little misunderstanding with his first girlfriend. He was reluctant to take the initiative to sort things out; rather, he was expecting her to act first. As a result, the two-year relationship was terminated by his pride. The male chauvinism that he upholds was even more explicit in his second relationship. When he was dating his future wife, he took every opportunity to find out if she was obedient. In this regard, the demand on women still existed as it did in Uncle Qiu's generation.

Before Zhichao was married, he found that his love for his future wife had disappeared. He thought of calling off the wedding. Personal feelings at this level were seldom found in Uncle Qiu and his peers. Compared with the traditional marriage, which was basically carried out by order of the parents, Zhichao obviously gave more thought to his feelings toward his wife. However, in the decisionmaking process, the tension between his personal feelings and traditional norms arose. To honor the traditional ethical responsibility in human relations, he chose to marry eventually.

The "virtues" of Zhichao's wife have won her a lot of praise from Zhichao and her in-laws. However, the heavy workload has transformed her body from petite to sturdy. Moreover, she did not care about her appearance. Compared to the beautiful and slender karaoke bargirls of the nineties, she looked like a clumsy old local woman. Zhichao frankly admitted that he wished he could have one or two "secret concubines," just like the well-off villagers. He also had sex with the female workers who came from other provinces. We can see that Zhichao adheres to the traditional value that "the woman stays with only one husband," and that a man can have other wives and concubines.

Because of the government's dedicated effort to spread the concept of freedom in marriage, it seems that Zhichao has learned how to resist his parents' interference with his marriage. He was conscious of autonomous choice on the issue of marriage. However, from his choice of a wife to his married life, we can see that he is not very different from his parents after all.

Since 1979, the reform and opening-up policy has improved the material lives of the people in the Pearl River Delta. People no longer follow the road of economic collectivization. They freely pursue various economic activities. The protagonist of this chapter, Qiming, grew up in this period. Upon leaving school, Qiming did not have to work in the fields. He was involved in economic activities led by the commodity economy. Compared to the older generation, the young people of Qiming's generation have more freedom of choice in the deployment of economic resources and in the way they live.

Not only that, the experience has broadened this generation's horizons. The traditional value system is no longer the only standard that guides their lives.

Let us take as an example the issue of choice in matters connected with marriage. Qiming chose to marry Meiling, who has the same surname. This choice touched on the taboo against endogamy, which was disapproved of by his parents and elders. Under pressure and despite feeling disturbed, Qiming insisted his relationship with Meiling was far more important than the traditional values upheld by his parents. His choice reflected his life experience under the reform and opening-up policy. He did not follow his parents' dictates. He was one step ahead of Zhichao. His relationship with Meiling was based on his choice.

After marriage, Qiming and his wife lived with his parents. The distinct differences in lifestyle between the two generations caused conflicts between the mother-in-law and the daughter-in-law. During the conflicts, Qiming stands by his wife most of the time. We see that his marriage presents a family pattern in which the "husband–wife" mode of association forms the core of a family and challenges the traditional family pattern in which the "father–son" mode of association forms the mainstay of the traditional familial ethical norm.

This shift could in theory slowly wipe out the traditional unequal relationship between the two sexes, of "men decide, women follow." However, the reform and opening-up policy has also brought about the widespread sex trade in the Pearl River Delta. Although Qiming values his relationship with his wife, he cannot resist the temptation of the bar girls. After marriage, despite the fact that he did not indulge in sex like the other villagers, nor go to karaoke or other sex establishments, he had sex with the bargirls without regret. In this regard, Qiming and Zhichao are much alike in the sense that they share the same traditional value of inequality between the two sexes. The sex trade that developed along with the reform and opening-up policy seems to help perpetuate the traditional value of inequality between the two sexes on yet another level.

Part II

Tradition, Women, and the Interpretation of the Self

7

Women in Chinese Villages

A Review of the Research

"In the home, follow your father; in marriage, follow your husband; in old age, follow your son." In the early 1980s, many Western social scientists, especially feminists going into the interior of China to do research on the living conditions of rural Chinese women, often used the concept of "patriarchy" to explain the structure of Chinese lineage and family.[1] From their point of view, Chinese women were dominated by the patriarchal culture in many spheres of activity, be they political, economic, or social. The proverb at the beginning of this chapter can be cited as evidence of the inequality between the sexes. However, from the point of view of Chinese tradition, the compliant behavior of village women reflects the ethical values of Chinese traditional culture in which the "family" is dominant. Women interpret the meaning of their existence in light of a male-dominated familial ethical order. Obviously, this differs greatly from the view of individual freedom cherished by modern Western societies. Feminists are aware of the differences between the two cultures, and in the explication of their research framework, they draw the reader's attention to the key characteristics of Chinese culture. Despite that, in the interpretation and analysis of the data, they appear to have unconsciously placed too much emphasis on the criteria of Western individual freedom. They see the compliant behavior of village women in a one-sided way—as a phenomenon of women being oppressed by patriarchal culture.[2] We believe that the research on Chinese women done in the mid-1980s is influenced by the Western women's liberation movement, then at its peak.[3] Feminists used Western doctrine to interpret and criticize the cultures of other societies. This reliance prevented a clear understanding of Chinese indigenous culture.

Since the end of the 1980s, there has been more discussion in the West of multiculturalism. More and more academics come to realize that the experience of modernization in the West is situated in the context of its own historical past and its present reality. It cannot be directly applied to understand and interpret the cultures of other societies. Scholars also emphasize the importance of respecting the uniqueness of other cultures. Under the influence of this new thinking, feminists have adjusted the direction of their research.[4] Among the issues under discussion is how, while respecting the traditional values of other cultures, one does not have to retreat to relativism and give up all cultural critique. A more central question is this: How does one appropriately apply the concept of liberation, an important insight advanced by Western feminist movement, to non-Western women's studies?[5] In order to answer these questions, it is necessary to engage in long-term theoretical exploration and substantial empirical research. The available literature shows, however, that the study of rural China in the 1990s has shifted its emphasis to the relationship between the countryside and the state.[6] Since 1990, the number of research studies on rural Chinese women is diminishing,[7] but their discussions are less biased than those of the 1980s. In this research, we have tried to combine the Western concept of "autonomy" and some of the salient characteristics of Chinese familial culture, and with a sympathetic viewpoint, explore some of the changes experienced by rural Chinese women in the era of reform and is opening-up policy. The focus of the research is on the rise of self-awareness of Chinese women. We will explain this in the following paragraphs.

To focus our research on the period of reform and opening-up policy gives our work a certain contemporary significance. Many commentators have pointed out that the development of Chinese society after 1979 differs markedly from the collective period of socialism between 1949 and 1978. It also differs markedly from traditional Chinese society before the Communists assumed political power. A broad look at history shows that after the decade-long Cultural Revolution, the Communists reduced their overemphasis on political ideological education and turned to the practical task of economic reconstruction. The Chinese government put into effect a series of reforms in the villages. It disbanded the people's commune and replaced it with the Household Responsibility System in agricultural production. From 1984, villages and towns began to develop business enterprises. The mode of production was gradually transformed from collectivization to individualized market economy. Industry and commerce gradually replaced farming as the village's main form of economic production. This change is particularly noticeable in the southern coastal region, where there are industrial and commercial enterprises. The rise of industry, be it governmental, private, or joint foreign and Chinese investments, created a demand for a large force of laborers, technicians, and other professionals. In 1979, the Chinese government abolished the system of the centralized assignment of jobs by the state. People were free to choose their own line of work. Many richer areas along the China's southern coast were able to implement a policy of nine-year compulsory education. These measures encouraged the mobility of labor among villages, towns, and even provinces. The inflow of foreign capital brought with it new technology, management skills, and information.

From the official point of view, the development appeared to prove the validity of the central government's change of policy—the revision of the extreme leftist route before 1976 and a new workable model of modernizing the village economy. There are a great deal of official survey statistics documenting the experience of economic reform in the countryside. The research reports produced in China also analyze the reasons for the success or failure of the economic reform in the village. Such studies show a shift toward adopting economics rather than politics as the basis for policy. From another angle, such economic developments are of great social import. Part I of this book has explicated the influence of the reform and opening-up policy on concepts about marriage. We studied three men—one old, one middle-aged, and one young—and examined their views toward choosing a wife, marriage, and marital life. We witnessed the transition from family-arranged marriages to those based on romantic love. In the case of Qiming, we see clearly that there is greater awareness of the option of choosing one's own spouse than was true in previous generations. To broaden the scope of discussion, we can consider self-choice consciousness as a part of consciousness of selfhood. We can say that the significance of the reform and opening-up policy lies not only in the narrow view of economic betterment as conceived by the government. In the broader sphere of social life, they have also promoted the consciousness of individual autonomy. In Part II, we stay within the time frame of the reform and opening-up period and, following the theme of the growth of individual autonomy, discuss the impact of reform and the opening-up policy on women in the countryside.

When we adopt the concept of autonomy to explore the condition of women in villages during the reform and opening-up era, we inevitably encounter the problem of the appropriateness of such a research orientation. Without a doubt, the notion of autonomy originates from the West. If we simply impose the concept onto the research into Chinese women without caution, then it is easy to end up with biased understanding, analysis, and criticism. This is exactly why we have reservations about 1980s Western feminist studies about Chinese women. The problem we encounter is similar to that faced by Western feminists in the 1990s. How do we come up with appropriate understanding, analysis, and critical remarks while adopting Western concepts? Although we are not able to provide a lucid and comprehensive answer, we have worked out a preliminary direction. First, we start from the standpoint of Chinese women in the village, and allow the subjects' own cognition to guide us toward a definition of "autonomy." Second, we take seriously the indigenous traditional culture of China, and do not treat it merely as a target for criticism. We take the ethical values of lineage as an important resource for understanding rural women.

During our fieldwork, we met women of different ages and living conditions. Our topics touched on how they deal with everyday routines involving clothing, food, housing, transportation, employment, marriage, and having children. It is obvious that such basic human needs are universal. They have to be dealt with, whatever the social and cultural context. On the other hand, how women understand and interpret the meaning of these needs will vary according to their different social and cultural backgrounds. Let us look at one obvious example. In the Chinese village, women feel it is their duty to give birth to a male child. One reason is that the traditional rural economy demands a large amount of labor. In addition, Chinese familial culture regards the male as the descendant of the family. However, to a woman living in a modern Western society, whether or not to have children is a decision made with a partner, and is sometimes a decision made alone. Also, the gender of the child is not generally a matter of concern.

When a woman in a Chinese village is asked about her attitude toward having children, she might have to cite a characteristic of Chinese familial culture to back her own view, so as to make the questioner understand the rationale for the desire to bear a boy. In answering, the woman will employ words and phrases which, to a greater or lesser extent, reveal whether or not she identifies with the values familial culture imposed on women. In other words, such replies will reveal, to a certain extent, the judgment Chinese rural women make of the cultural context in which they find themselves. Of course, such judgments are rather subjective, since they are based mainly on the speaker's views and feelings. This may not be a bad thing; it may strengthen the validity of the present study. On one hand, since our study is about Chinese rural women's consciousness of the self, their subjective views and feelings are important data for our study. On the other hand, these subjective views and feelings are not vacuous or unrelated to real issues. They are anchored in the social and cultural condition of the subjects. We think that since this part of the research studies Chinese rural women's consciousness of the self by seeing how women understand and interpret self-fulfillment, we can, to some extent, employ a concept developed in the West and apply it to the Chinese subjects we interview in our research.

We need to have, however, a clearer definition of Chinese society and culture in our interpretation and analysis of the fieldwork data. We have concentrated on the period of the reform and opening-up policy for our study. The main feature during that time, as indicated earlier, is that industry and commerce moved into the village on a large scale, bringing about huge unprecedented changes in the economy and social life of the village. Some scholars have postulated that although from 1949 to 1976 the Communist Party had tried to destroy traditional Chinese culture by imposing a series of political movements, their success is still a matter of debate. When the reform and opening-up policy were introduced in 1979, they brought about a market economy similar to the modern model, which allowed the emergence of individualized economic activities, as well as individual wealth outside of what is inherited from the family. These changes might be taking place at the more basic level of peasants' lives and undermining their identification with traditional family culture. Because of this, we will consider the reform and opening-up policy as a change from tradition to modernization.

This assumption helps us to place the change in Chinese rural women within the broader canvas of social change. We can reformulate the question like this: What is meant by self-concept for Chinese rural women living in a transitional period between tradition and modernity? Answering this question presupposes our understanding of the self-concept of a modern person. Charles Taylor systematically examined Western culture from the ancient Greeks on for over 2,000 years and sketched out the concept of self in the West as it was molded by the cultural currents of the various periods.[8] One thing became obvious. The ideas of "human rights," "freedom," "fairness," and "private property" are values that developed from the historical-cultural tradition of the West.

We can say from the experience of the West that the emergence and development of "modernization" cannot be transplanted from the outside. It must evolve from the indigenous experience of the traditional culture. Therefore, when we view the development of self-concept in Chinese rural women, we cannot ignore traditional Chinese culture, in particular how familial culture has molded rural women's traditional self-concept. The ethical values of the family such as lineage descendency, obedience, loyalty, and faithfulness are traditionally important components of women's selfhood. We will portray how rural women's understanding and interpretation of self-fulfillment gradually shift from the family and lineage to individual wishes.

In chapter 8, we present the experience of a young village woman growing up and observe how she understands and interprets the need for self-fulfillment in the social environment of rural economic reform. Chapter 9 discusses the lives of an old woman and a middle-aged woman in the village. Through their identification with traditional values, we hope to show by contrast the change in self-concept of the young woman in the previous chapter. We must point out, though, that old and middle-aged women were also living in a village undergoing rural economic reform and therefore subject to the impact of new social phenomena and new values. The difference is that they grew up, respectively, in traditional Chinese society before the Communist rule, and in the socialist collectivization period between 1949 and 1978. Of course, the two women's degrees of identification with traditional familial values are different in important ways. We hope that by looking at three rural Chinese women—one old, one middle-aged, and one young—we can capture some of the features of the development of women's self-concept.

8

Wang Guizhen

Growing Up Under the Reform and
Opening-Up Policy

It was an evening in January 1995, after the winter solstice. The weather was getting cold in Baixiu Village. At around six o'clock, the sky clouded over and the temperature dropped to 13 degrees Celsius. Although I was having a hot dinner at the research center, I could still feel the cold wind blowing in through the wooden front doors. Normally, I closed the wooden doors when I had dinner. Tonight, since I was going to have an interview with Wang Guizhen, a female villager, I left the doors open to welcome her.

Wang Guizhen, who is usually called Ah Zhen, was born in 1968, at the time of the Cultural Revolution. Since Ah Zhen was just a child at that time, the rise and fall of the political struggles hardly left their mark on her. What really affected her growth was the reform and opening-up policy immediately following the Cultural Revolution. This began in 1978, when Ah Zhen was ten years old. In December 1978, the central government adopted the reform and opening-up policy of the countryside, and that transformed China's rural areas. Collectivization of the commune was replaced by the Household Responsibility System in Agricultural Production. In September 1979, the central government proposed the development of commune team enterprises (renamed "village–town enterprises" in 1984).[1] Many Hong Kong and Macau businessmen built factories in the villages on the coast of Guangdong Province.[2] The importance of education in China was also reaffirmed after the ten-year Cultural

Revolution (1966 to 1976).[3] In 1979 and 1980, the central government officially implemented the single-child family policy, which had been planned for years.[4]

Ah Zhen is twenty-six now. Like other village women, she assisted her parents in working the fields and did the housework in her childhood and early youth and hoped to get married at the right marriageable age and have children. However, the rural economic reform coupled with the other social measures afforded her an opportunity to attend junior and senior middle school. Having graduated from senior middle school in 1988, she still lived with her family and was registered in the village as a resident. Yet she did not have to work in the fields; instead, she worked in a factory with investment from Hong Kong.

Ah Zhen can be taken as a typical example of the young women growing up under the rural economic reform policy of "leaving the soil but not the village." Compared with women of the older generation, she obviously has a broader perspective. For example, apart from sustenance, she looks for enjoyment and a lifestyle that suits her. On the one hand, she tries to respect the wishes of her family. On the other hand, she attaches great importance to her own ideals and interests, sometimes even questioning the values upheld by the traditional village women. The growth of business and industries offered opportunities that were never available before. The reform policy fundamentally changed the social environment of Baixiu Village, and economic development helped to elevate individual self-awareness. When these types of changes occur, the younger generation starts to question the traditional value of working for familial ideals as one's own goal. My own view is that Ah Zhen's experience reflects the changes in self-understanding that a village woman undergoes as a result of the reform and opening-up policy, in particular how she understands her own needs for fulfillment, plans for the future, and faces limitations when she makes choices.

While I was having dinner, I heard a roaring sound outside. I put down the chopsticks and went to the front door to check it out. It was Ah Zhen riding her new motorcycle past the gate into the front yard of the research center.

The soft yellow light from my living room shone on Ah Zhen and her red-and-black motorcycle. Its style and model were seldom seen in this neighborhood. Ah Zhen's diminutive stature did not match the height of her motorcycle. When she dismounted, she had to incline her body to the right so that her right foot could touch the ground. She looked proud and smiled contentedly.

I invited Ah Zhen to come into the living room, but she anxiously asked me where to park her motorcycle. "You know, it's not safe here and my motorcycle is brand new!" She drew her eyebrows close.

I looked around the house and suggested moving her motorcycle inside. However, as we tried to push it in, we found that there was a fifteen-centimeter threshold right in front of the doors of the research center, which was an old building. We were not able to raise the motorcycle above the threshold. After some discussion, we decided to leave it outside. We kept the doors open so that we could see the front part of the motorcycle right next to the door from the living room while we were talking. Though it would be cold, it would make Ah Zhen feel comfortable during the interview.

After settling the motorcycle issue, Ah Zhen sat down with contentment. She looked at the food on the table and said, "You haven't finished dinner yet."

I looked at my watch and said, "Our appointment is at seven. I didn't expect you to come half an hour early."

"I had my dinner in the canteen, so I came early. If we finish soon, I can go home to watch television." Ah Zhen smiled.

By 1980, only a few families had black-and-white television sets. Thanks to the economic development over the next twenty years, villagers became wealthier. Now, most of the people in Baixiu Village have color television sets, which sell for between 1,200 and 3,000 yuan. Even though the China Central Television broadcasts in Putonghua, and the Guangdong Station broadcasts in Cantonese, the villagers prefer to watch Hong Kong programs. Around 9:00 every night, you can hear the theme songs of the Hong Kong television series whether you are walking on the streets or strolling through the alleys between the houses. A villager once told me that before the rural economic reform, villagers mostly got together in the evenings to chat or hold meetings. However, the end of the commune system and the popularity of television combined have resulted in villagers' usually staying at home to watch television. For some housewives who do not go out to work, watching television has become a major daily activity in addition to doing the housework, taking care of the children, and chatting with neighbors.

Because of her long hours at the factory, Ah Zhen does not spend much time watching television. The Xingsheng Electronics Factory where she works is just like other factories. She has to work overtime five days a week. On those days, she works from eight in the morning to nine in the evening and has one and a half hours for a lunch break and one hour for a dinner break. Like the other workers, she eats in the canteen. The only days she does not work overtime are Wednesdays and Sundays, when she leaves at half past five. She has only one Sunday off every two weeks. I make the following calculations: after taking away meal time, Ah Zhen works twenty-eight days a month on average, nine hours and thirty-four minutes a day, and sixty-seven hours a week.

Ah Zhen is stressed out by the long hours. On her time off, if she does not go downtown, she will sleep all day. No wonder she has black circles around her eyes. There were a few times when I visited her around nine in the evening and the following scene would occur. I would find Ah Zhen and her two sisters, who also work in the factories, lying under quilts on the couches watching television. As soon as they saw me, they got up, fixed their hair, straightened their quilts, and sat up to talk to me. After we got to know each other better, Ah Zhen would stay put and chat with me.

In order not to disturb Ah Zhen, I interviewed her only once on a Sunday afternoon. We usually had interviews on Wednesday evenings and finished before nine so that she could be home in time to watch television.

Tonight, she arrived at the research center half an hour early. She said since she already had dinner and had nothing to do, she might as well come earlier. Evidently, she does not go home for dinner even if she does not have to work overtime. I found this strange because dinner at home is usually better than the "big wok meal" in the canteen. It appears she has dinner with her family biweekly. She does not feel bad about this way of village life.

Rural Life

Ah Zhen told me about her childhood in the village. From the age of five, she helped her parents in the fields. In around 1974, the land in Baixiu Village was mainly for farming. Factories hardly existed. Most of the inhabitants were locally born farmers. That was the period when socialist collectivization economy was practiced.

Ah Zhen detailed the hardship of the rural life: "When I was five, my younger sister Nianzi was only three and the second sister Fenkai just a baby. I already had to deliver meals to mom and dad in the field. At that time, I was just a kid, but grandma had already put a pole on my shoulder to carry a container of food on the left end and a container of soup on the right end. It took me almost an hour to walk from the house to the field. What bothered me most were the bridges, which were very muddy. On rainy days, they were even worse. I had to watch out, otherwise I would slip."

Ah Zhen stared at me and said, "You know, there's a river below. If you fall, you'll be drowned. Thinking back, I'm still a bit scared.

"We cultivated, including transplanting, reaping, planting, and so on. We planted bamboo grass, which is used to make straw mats that you sleep on in summer. We cut the stalks in half and laid them on the ground to dry in the sun till they were ready to be knitted." Ah Zhen paused and said, "Farming is hard. The grass is heavy. We needed about 100 catties. Can you imagine small kids like us having to carry the grass? It's really heavy. Mom had to carry it too!

"Girls transplant rice seedlings in the field when they are five or six. We stand in the water and bend down but we're fast. When we're ten, we pull up rice seedlings and we're fast too!" Ah Zhen smiled. It looked as if in her memory she were competing with someone.

In the picture portrayed by Ah Zhen, it seemed as if there were no boys. Out of curiosity, I asked, "Well, where are the boys? Why didn't they have to work in the field?"

"I don't know why. Boys are slow learners, especially about planting seedlings." Ah Zhen spoke proudly, "Since we girls learn fast, we mostly work in the field. However, if there are only boys in a family, they have to go too. Yet, they are not as good as we are. If there are boys and girls, the girls are faster, so they go to work in the field. The boys will only go when they are fifteen or sixteen. They are stronger so they can plough."

"Isn't that a bit unfair?" I said jokingly.

Ah Zhen smiled and explained to me in a serious tone, "I don't think it's a question of inequality. Imagine, those boys don't know how to plant seedlings. What can they do? Can you ask a five-or six-year-old boy to plough? How can you? Only when they are fifteen or sixteen can they do it. I think it's better for girls to do planting and boys ploughing. It's division of labor. I don't think it's a question of inequality," she affirmed.

Evidently, Ah Zhen was happy with what she and other girls had accomplished in the field. She spoke with confidence and flaunted her abilities. I let her go on about her farming experience as a child, then changed the subject to family planning, attempting to check the effects of the government policies and traditional values on the status of village women.

Family Planning

Since the early 1950s, Chinese scholars like Ma Renchu have raised the issue that excessive growth of population will obstruct the economic development of China.[5] In 1956, China launched the first family planning campaign and distributed directions for contraception. However, not long after, Mao Zedong launched the Great Leap Forward in the late 1950s. The original birth control policy was changed to a policy encouraging reproduction, thus spurring population growth. In 1962, the Great Leap Forward was over and stability was restored. Family planning was implemented for the second time, but in the mid-1960s, the central government waged the Cultural Revolution. Family planning was brought to a halt by the continuous political struggles. According to statistics, in the twenty-one years from 1949 to 1970, China's population grew from 542 million to about 830 million[6] at a growth rate of approximately 53 percent. The Cultural Revolution died down in the mid-1970s and family planning was revived. In 1972, the State Department started to spread the ideas of late marriage, delay in giving birth, and birth control in the cities and rural areas. It was stipulated that the marriage age for village women should not be younger than twenty-three, and twenty-five for men. The first child should be at least three years older than the second child. Each couple in the rural areas could have a maximum of three children, and those in the cities a maximum of two. In 1977, this was reduced to two children in the rural areas. Along with the reform and opening-up policy implemented in 1979, the central government announced the single-child policy for the first time. In the rural areas, however, the peasants were allowed to have two children. Since then, there has not been a nationwide political movement, and family planning has been practiced up to now.[7]

Ah Zhen told me that family planning was not practiced in Baixiu Village until 1972. Before then, the policy was only propaganda and was not carried out thoroughly. Her mother gave birth to three daughters before that year: Guizhen in 1968, Nianzi in 1970, and Fenkai in 1972. Evidently, in the year of Fenkai's birth, the central government still allowed couples in the rural area to have three children but not four. In other words, the births of Ah Zhen, Nianzi, and Fenkai were not affected by the birth control policy.

During home visits, I also met Ah Zhen's two younger sisters. Nianzi (literally "expect son") is now twenty-four, and is slim and taller than Ah Zhen. She also has more acne than Ah Zhen. She dropped out of school after she finished junior secondary. She speaks with a strong local accent. She works in a knitting factory in the next village. Whenever I visit, she is always lying under a quilt on the couch watching television. Whenever she sees me, she serves me oranges, dried longyan, or peanuts. She is not talkative. When Ah Zhen and Fenkai come home, she sits quietly watching television or listening to us.

Nianzi's younger sister, Fenkai (meaning "divide"), is twenty-two. She is tall and has short, straight hair. She and Ah Zhen look alike. Having graduated from secondary school, Fenkai has been working in an electronics factory in Baixiu Village. Unlike Ah Zhen, Fenkai found the work of a foreman monotonous and inflexible, and the hours long, so she chose to be a customs clerk. Her work required her to travel between the ports of

Shenzhen, Chaojiang, and Juetou. Perhaps it was the economic reforms in the past ten years that contributed to the rapid development of the Pearl River Delta and the reduction of farmland. In a lot of places, in particular the cities and developing towns, road construction, factory building, and housing construction have polluted the air. Fenkai told me with annoyance that since she was out there every day, acne appeared on her face. Nianzi inadvertently disclosed that Fenkai also moonlighted doing customs work for the other factories in the hope of earning more money. Despite the government's limits on the number of children allowed, Ah Zhen's parents had a fourth child: an eighteen-year-old son, Tamking. He studies senior secondary in a school in Qingyang Town. He is a boarding student, so he comes home only on the weekend.

Fenkai is less talkative than Ah Zhen. Talking to Ah Zhen is rather enjoyable. She is more straightforward. When she is engaged in the conversation, she becomes very animated, using gestures to emphasize her meaning. But it is not easy to talk to Fenkai. She rolls her eyes, thinks for a while, then speaks slowly. Perhaps she is cautious because of my identity as a researcher. Whenever I ask for her personal opinion, she will try to generalize, so it is difficult for me to know what is on her mind.

One evening, I went to the Wangs' house for an interview and Ah Zhen was not in. I chatted with Fenkai. Unexpectedly, she volunteered that she was thinking of changing her name. "'Fenkai,' what good is this name 'Fenkai'! A lot of girls in the village are called 'Fenkai.' What good is it! I want to change my name, but I have no idea what a good name would be."

Actually, when I first heard of the name Fenkai, I found it strange, yet I had no idea that she disliked her name that much. In addition, it had never occurred to me that a girl growing up in a village would change the name that her father gave her. Is there some underlying meaning? I guessed that perhaps Fenkai was subconsciously trying to express her discontent toward the low status of women in the familial culture. In this regard, perhaps I might find some clues from my interview with Ah Zhen tonight.

Ah Zhen and I were still on the topic of family planning in Baixiu Village. We had just talked about the family planning mandate that was carried out in Baixiu Village in 1972. The central government ordered that couples in rural areas could have only three children. I asked Ah Zhen how they would be punished if they gave birth to a fourth child. She said they would become "blacklisted" and not eligible for the rations of food. Mrs. Wang did not express an opinion, but her husband had insisted on having a son. Ah Zhen explained, "My dad is the only son. My grandpa's eldest brother has three girls, his second brother has no children, and his third brother is not married. So, my dad wanted my mom to give him a son."

Having heard that the third generation is all girls, I said casually, "No doubt about it, your family can be called 'a family full of girls!'"

"Oh yes! Many girls, but the family won't strike it rich!"

"What does that mean?" I asked.

"All in all, sons will bring you wealth but not girls!" Ah Zhen said with discontent and did not give me a detailed explanation like before.

I did not expect this reaction, so I tried to change the subject. I asked her how her parents reacted to the birth of the first child. Ah Zhen paused and started to smile and said, "When I was born, there was not a special celebration, but everybody was happy and treated me like a princess. So, they called me Guizhen because it carried the meaning of preciousness. The second baby was called Nianzi, which meant they wanted a son. The third baby was also a girl, so they called her 'Fenkai.' Hopefully, she would be the last girl and bring a son. Finally, the fourth baby was a son."

I was a bit puzzled. Ah Zhen had mentioned earlier that in compliance with the orders of the central government, giving birth to a fourth child was prohibited in Baixiu Village in 1972. In that case, how could Mrs. Wang give birth to her youngest son without interference?

Ah Zhen explained that the local government was not consistent with the central government in executing the orders. She recalled, "Though it was prohibited to give birth to a fourth child, the order was not obeyed strictly. It was not until 1974 that the government was serious about it. Afterward, the policy was sometimes lax and sometimes tight. I remember it was most stringent in early 1976. Yet by the end of 1976, it was lax again. Next year, that is early 1977, it was tightened up again. My brother was born at the end of 1976. He was lucky!"

When she mentioned those days at the end of 1976, Ah Zhen looked nervous.

Normally, childbearing is the concern of the adults. An eight-year-old girl should care about playing and working in the fields, as we discussed before. Yet Ah Zhen remembered well the chronology of the family planning policy and how stringent the policy was. Evidently, everyone in the family must have been very nervous during the year of 1976 when Mrs. Wang was carrying the baby. They must have worried about whether Mrs. Wang's pregnancy would be discovered by the work team in the commune. So even the little Ah Zhen knew of the details of the family planning policy.

Obviously, Ah Zhen is pleased that her mother took advantage of the lax policy and successfully gave birth to a younger brother. She is like her father in the sense that she was anxiously expecting the birth of a brother so as to continue the family lineage. It is the typical traditional rural culture.

Perhaps it is because Ah Zhen was the first child that everybody values her even though she is a girl. She was named "Guizhen," meaning "highly placed, precious," which reflected her place in her father's heart. However, Nianzi and Fenkai were a bit different. Their names reflected the fact that that they were viewed as placeholders while the family awaited the birth of a son. The names also reflect the low status of women in traditional culture. When she grew up, the more independent Fenkai started to entertain the idea of changing her name.

Since the birth of Takming, Ah Zhen's father was relieved of the worry about a lack of male descendants. He had fulfilled his duty. Mrs. Wang was also relieved. Shortly after, Mr. Wang went to work in Hong Kong. He came back four or five times a year to see his family. Mrs. Wang stayed in the village with her four children.

Ah Zhen's father was not home for long periods of time, and her mother was mild and feeble and easily bullied. Being the eldest daughter, Ah Zhen was well aware of the fact that she had to take care of her younger siblings and assist her mother with the housework. Gradually, she trained herself to be strong and decisive. Sometimes I think that her will and ability are well above some men's. In particular, this is reflected in the way she pursued her education.

Pursuing Education

Since the founding of the People's Republic, the Chinese government has actively modeled itself on the Soviet education system and attached great importance to knowledge and skills training in order to build up the economy.[8] Yet, in less than ten years, in 1958, the central government launched the Great Leap Forward, which emphasized the development of the country by political mobilization. The original aim of the education policy was shifted from serving economic development to nurturing the political consciousness of the people. In 1962, fault was found with the Great Leap Forward. In response, the central government hastily made adjustments by reinforcing education aimed at knowledge and skills training. However, four years later, the Cultural Revolution began. The whole society was involved in a class struggle in which education once again became a tool for instilling political ideas. The ten-year Cultural Revolution ended in 1976. The central government attempted to make amends for the education that was lost in the Cultural Revolution by re-emphasizing knowledge and skills training. They trained people to meet the needs of developing technology, industry, agriculture, and the military. Since then, China has entered the Four Modernizations period. In April 1978, the National Education Work Conference preliminarily prepared the draft outline of the objectives of a national education policy for the period from 1978 to 1985. In May 1985, the Decision of the Central Committee of the Chinese Communist Party on Educational Reform was released, making the local government responsible for the development of basic education. A planned nine-year compulsory education was to be implemented accordingly. At the same time, it was confirmed that education must serve the cause of socialist construction.

According to Ah Zhen, it was 1976 when she went to primary school. At that time, she was eight years old. The Cultural Revolution had just ended and the continuous political movements also died down. She received six years' primary education in a stable political environment in Baixiu Village. When she was thirteen, she was admitted to a secondary school in Qingyang Town where she did her junior secondary and finished her nine-year basic education. In 1985, she continued her schooling by enrolling in a senior secondary school in Qingyang Town. Three years later, in 1988, she graduated and returned to Baixiu Village to live with her family. At that time, her father was already retired and came back home. She worked in a Hong Kong–owned electronics factory as a clerk.

In terms of the educational development in China, Ah Zhen's education went hand-in-hand with the Four Modernizations. Compared with other women in the rural areas, her generation obviously received education in an environment full of opportunities. Nevertheless, we all know that the traditional farming society does not encourage women to study. Ah Zhen had to break the traditional value of "no knowledge is best for women" to achieve her academic qualifications. As a result, I am interested in listening to her description of how she studied. Perhaps from this, I can find out how the women of the new generation raised their self-awareness.

Ah Zhen stared at me and it seemed as if something had suddenly come to her mind; then she smiled. She recalled how she got into the secondary school when she was thirteen. "In 1982, I was admitted to the secondary school. That was the first year that children in the rural areas had the opportunity to go to study in Qingyang Town. That year, there were sixty candidates and only ten succeeded in studying in Qingyang Town and I was one of them. There were four boys and six girls, all with an average score of seventy, but I got eighty." Ah Zhen smiled radiantly.

Actually, apart from her personal abilities, the objective environment of Baixiu Village also facilitated her continuing her education. As early as 1985, China had already decreed the expansion of basic education. However, resources in the rural areas varied. Not all of the children in the inland villages had the chance to finish basic education. It was even worse for the girls. The traditional value of superiority of boys over girls deprived girls of an equal opportunity in education. However, the situation in the Pearl River Delta in southern China was quite different. The reform and opening-up policy implemented in 1979 encouraged the rural areas to develop village–town enterprises. Baixiu Village and the nearby areas thus prospered and this provided the rural areas with the objective conditions to implement basic education. As Ah Zhen mentioned, secondary education was open to all children of her generation and had good results in Baixiu Village.

Ah Zhen had done quite well in the primary school. She told me it never occurred to her that she could fail to gain entry to the secondary school in town. Even when she applied for the senior secondary at sixteen, she was very confident.

"Compared to the junior secondary, I was more nervous when I sat for the senior secondary exam. The center for the junior secondary entrance exam was at the nearby Xijing Village. At that time, I was one of the top ten in class and the teacher would select me as a school representative to sit for the Regional Centralized Examination. So, I was not worried. Yet, I was more nervous when I took the entrance exam for the senior secondary. I was a boarding student, so I usually went home to see Mom on Sundays. However, during the period when I was preparing for the exam, I did not go home on Sunday, rather I studied in school."

Ah Zhen paused and seemed to reminisce about those days of hard work. Then she continued, "When I was studying secondary form three, there were four classes and two of them were the elite classes. I was in an elite class. So I was very confident. The only thing that bothered me was which senior secondary school I was going to."

I had the impression that few girls in Baixiu Village finished senior secondary schooling. Obviously, Ah Zhen belonged to the minority. It was not only her academic ability that enabled her to gain entry to the senior high, but also her luck.

Ah Zhen's youngest sister, Fenkai, once told me that there were some girls in the village who were intelligent and did very well at school, but they dropped out of school after they had finished junior secondary. It was not because they failed in the competitive exam; rather, they were held back by their families. Ah Zhen's father did not stop her from taking the examination, but he did not give her any encouragement either. He was noncommittal: "If you are admitted, go ahead!"

Although her father was not very supportive, he left her room to make her own decision. Since secondary school education was not common in the village, Ah Zhen thought hard before she made the decision.

"At that time I thought, why couldn't I go on to senior secondary after I've finished junior secondary? I also thought, having finished senior secondary, can I go on to post-secondary or university? The reason that I wanted to study senior secondary was to know my standard and to test my ability so as to find out my strengths. A career in banking? Department stores? Shops? Others?" Ah Zhen paused and said, "I hoped I knew how to make a choice."

Ah Zhen emphasized that she knew very well why she wanted to study senior secondary. On the one hand, she wanted to know what her strengths were. On the other hand, the qualification offered her more choices in her career.

I think this is very important in understanding the transformation of village women. For the women before the rural economic reform, the agriculture-oriented village environment provided a set lifestyle. From childhood to adolescence, youth to marriage, they mainly work in the fields and do housework. Change is unlikely, so they do not give it any thought. However, Ah Zhen is of the new generation. Although she lives in the village, she is not satisfied with merely surviving, as the traditional farmers do. The rapid changes in the rural environment, the flourishing businesses and industries, the improvement of the transport system between towns and villages, and her academic qualifications have all made an impact on her. In her opinion, a "career" is not only a bread-winning tool, but comprises a job selection process and practice that help to enhance one's self-understanding.

Working in the Village

In 1988, when Ah Zhen graduated from secondary school, there were normally three career choices. The first one was banking, where you could either work the morning shift or the afternoon shift. One would have more free time but the salary was low, around 300 yuan per month. The second option was to work at the state-operated shops as a salesperson for a monthly wage of about 400 yuan. The third choice was to work at a factory, where the monthly wage could be as high as 400 to 500 yuan. Men had better prospects than women. They usually worked as customs clerks or in accounting. Those who had *guanxi* (connections) would become factory managers while those who were more fortunate would inherit the family business.

Ah Zhen recalled her decision-making process and said, "At first I wanted to work in a bank because I would have more free time. But then I thought about what I would do after work. I still had half a day. It would be very boring if you had too much time. So, I chose to work in a factory because the pay was higher."

I thought that from the financial point of view, working half a day in a bank and taking a part-time job would earn more money than working in a factory. However, Ah Zhen said, "Despite that, if I work in a bank, transportation expenses will take up part of my salary since I have to travel between Baixiu Village and Qingyang Town. Actually, I have a classmate who works in a bank in the morning, as a salesperson in the afternoon, and has a part-time job in the evening."

Suddenly, Ah Zhen raised her voice and said, "But she lives in Qingyang Town and is a local and registered as a city resident. She has connections with 'those guys!'"

It seems to me that because Ah Zhen was not a local resident of Qingyang Town and does not have good connections with the people there, she decided to go back to the village. After a while, she spoke in a low and deep voice, "Oh, I thought since I had been in Qingyang Town for a couple of years, I needed a change. It's time for me to go back to work in a factory in Baixiu Village." This was the reasonable explanation she gave herself.

Shortly after she went back home, Ah Zhen found a job in Xingsheng Electronics Factory, an enterprise with Hong Kong investment. The factory complex comprised six three-story buildings joined by overhead corridors and had an area of 6,500 square meters. It was founded ten years ago and is the biggest of the dozens of factories in the village. Baixiu Village only had a population of about 2,000, so it could hardly provide sufficient labor for the factories. As a result, workers from other provinces were drawn to it. In ten years, the population grew to more than 20,000, mainly women. Xingsheng Electronics Factory had a staff of a few thousand and most of them came from Sichuan and Hunan provinces. Female workers between the ages of sixteen and twenty-four were in the majority. At first, Ah Zhen was only a worker. Equipped with higher academic qualifications and being a local, she was promoted to work in the warehouse six months later, to foreman II nine months later, and to foreman III in purchasing two years later. All in all, she has been working with Xingsheng Electronics Factory for seven years.

Once Ah Zhen showed me around the factory and led me to the Number 1 building. It was filled with an incessant, extremely high-frequency noise "Wa! Hu! Wa! Hu!" I had to raise my voice to speak to her. She told me it was where the sound boxes were manufactured. Hundreds of non-local female workers sat at the production line in rows. These girls who once worked in the fields worked diligently, skillfully, and rapidly. Yet they all had the same expression and silently worked with the parts in front of them for fear they would be taken to task if the work piled up.

When we walked to the other side, a strong nauseating smell came from soldering tin. I told Ah Zhen, "That smell is very irritating!"

"I'm used to it!" Ah Zhen said.

When she first joined Xingsheng Electronics Factory, she had to start by working on the production line even though she had a secondary school diploma.

Then we went to the warehouse and it was completely different from the production building. There were seven desks in the eighteen-square-meter office. Apart from the two electric fans mounted on the wall, there were posters of Hong Kong singers such as Liu Dehua and Li Ming. Some female workers were reading fiction or knitting sweaters and some were joking and forming "a small noisy market" in groups of three to four. As soon as the products were delivered to the warehouse, the group head would order the girls to receive and check the goods. Having finished, the girls would return to their seats and continue what they had been doing before. Some of the girls here worked with Ah Zhen before she was promoted. After chatting with them for a while, Ah Zhen led me to her current office.

Ah Zhen worked with the middle and senior management. The 180-square-meter room was divided into offices. It was almost lunch time and most of the staff had gone out for lunch. Ah Zhen pointed at the desk at the edge of the room and told me, "This is where I work. Take a seat, please."

She leaned against the desk behind her and told me the computer technicians and most of the middle management were university graduates from the outside provinces. While she was speaking, an overweight man aged above forty came into the room. A local, he was called Wang Ah Sen, had short curly hair, and wore a white shirt and gray pants. Ah Zhen smiled and introduced him to me, "He's our factory director."

Ah Sen greeted me with a smile. He told me that his job required him to go to different places. Although he had been talking for a couple of minutes, it seemed as if he were beating around the bush. I could only learn that he had to go out a lot but nothing about his work.

There were four directors. Ah Zhen also introduced me to three others. They were tanned, stout, and looked like they had been farmers at one time. They gathered in an office in a corner of the room. There were cigarette packets and a few documents on the desks. It seemed as if there was not much work. Like Ah Sen, they were locals, middle-aged, and had good relations with their kinsmen and the party secretary of Baixiu Village.

Generally, both local and outside women were ordinary workers. Those with higher education would likely be promoted to the head of the production line or other higher positions, yet they would never become directors. Middle-aged women with little education would be responsible for cooking meals and cleaning the quarters of the non-local female workers. These are mostly part-time jobs. When they are not working in the factory, they work in the field.

There was a great disparity in the career prospects of men and women. No matter whether a man was capable or not, he would be promoted to a higher position based only on his "local" identity and *guanxi*. Women marry sooner or later. If they marry somebody in the next village, they lose their "local" identity. In a patriarchal culture, women are always on the sidelines.

Ah Zhen intentionally made friends with different people, including her supervisor, other supervisors, peers, staff of the factories, even the supervisors in the other towns. Once she took me to a town that was miles away from Baixiu Village and purposely introduced me to a factory manager who was "close" to her. From my observation, this "close" manager was rather cool to her.

Ah Zhen kept telling me, "I like to make friends with different people. I don't care who they are. I'm approachable." Obviously, in a society where *guanxi* is a social capital, Ah Zhen is trying to build up her personal network; in her opinion, this was the "opportunity."

Finding a Partner and Marriage

The village–town enterprise attracted a lot of foreign female workers to Baixiu Village, resulting in a serious imbalance in the proportion of men to women in the population. The influx of the foreign female workers not only stimulated the growth of the sex trade, but also lessened local women's chances of finding a husband. Comparatively speaking, plain-looking local women were put in an even less favorable position. This problem has long been known to the villagers. However, because of the flourishing economy, the issue of prosperity had become the talk of the people. The marriage problem of the local women was treated merely as personal misfortune.

The twenty-six-year-old Ah Zhen was also bothered by the marriage problem. Her close schoolmates had all gotten married with the exception of one, who was a high school teacher. Some even had children, yet Ah Zhen did not even have a boyfriend. In her opinion, the out-of-province women were their major competitors.

Once I went window shopping with Ah Zhen in Qingyang Town. Like Baixiu Village, Qingyang Town was full of girls from the outside provinces. The only difference was that the girls there were all dressed up. The fashionably dressed women usually worked in karaoke nightclubs or the like. On the way, a non-local woman in a long chiffon dress and high heels brushed past us. Ah Zhen stared at her graceful movements and said, with a mixture of admiration and contempt, "They came from the other provinces! Quite pretty, aren't they?"

She then let out a low "Hum!" and took a glance at the woman's back, then moved on.

Having known Ah Zhen for a couple of months, I asked her to visit Dongguang City with me on a Sunday afternoon. We spent an afternoon in Ke Yuan, one of the four famous gardens of Guangdong. There was an out-of-province lady in ancient costume singing Cantonese opera. Her voice was clear and mellow. She sang both the male and female parts skillfully and got a big round of applause. What Ah Zhen liked most was the graceful manner in which she carried herself when she acted. Ah Zhen also praised her beauty. Ah Zhen listened to the lady with rapt attention. At the end of the opera, she seemed a bit upset. She cupped her chin in her hands and sighed, "Why are these outside girls so pretty?"

She mumbled this to herself, so I did not respond. After the performance, I took her to a cafe and we had some snacks. As usual, we chatted about everything. She was feeling better. When we mentioned the outside girls who worked with her, she appeared unhappy again and talked about marriage.

"Those who came from the outside provinces, especially the 'northern girls,' held onto the local guys. . . . I think, if I don't really like the guy, how can I marry him? Am I going to be like my mom who developed the relationship after marriage? That's impossible! He should finish secondary school; if not, at least he should have done his junior secondary. He should have a good personality and must speak the local dialect. I don't want to marry those coming from the other provinces since I won't be able to communicate with them."

Ah Zhen listed her criteria, which showed that she really wanted to get married. However, afterward, it appeared that she lacked confidence and was at a loss.

"I'm afraid that I might be too good for the guys. I'm too short and they might not like that. The older I am, the more difficult it is for me to find a guy. Most of them do not want to marry someone older than themselves." Ah Zhen was always bothered by her acne and husky voice.

"But I don't want to be like the other girls who always yield to the guys. I asked myself why I have to do that. I am not incapable and can support myself. There's no point to put up with all this. Even if I find a guy I like, I won't please him. Otherwise, I have to please him for the rest of my life after marriage. What's the point?" Ah Zhen looked displeased and once again showed an expression of stubbornness.

"So, sometimes I have thought: I might not be able to find someone! When I was down, or had too much work, I would ask myself, 'What's the point of living like this?'" Ah Zhen softened.

The Motorcycle

The marriage problem has always been on Ah Zhen's mind. She candidly admitted that she felt the pressure was greater in the village than in the city.

A woman needs a husband to settle her heart. As we have mentioned before, the reform and opening-up policy in 1979 changed the society. It was much simpler for the village women of the older generation to find a husband. Nowadays, it is more difficult. Comparatively speaking, women now need to give serious thought to marriage and make conscious choices. Ah Zhen is reluctant to just find somebody to marry, so she is still single.

Ah Zhen would only open up and pour out her worries and troubles accidentally, as when she met that beautiful outside lady in Ke Yuan whom she could not rival. Usually, she feels that she is a financially independent woman of the new generation. She seems carefree and confident. It is important for her to enjoy her leisure time. Whenever there is a long holiday, she will ask her friends to go on a trip, whether near or far. When she is off work, she will watch Hong Kong television programs. When she does not have to work overtime, she will go to Qingyang Town or the towns nearby for a stroll or to visit friends.

One Wednesday evening, I asked Ah Zhen to have dinner in a Hong Kong–invested restaurant in Qingyang Town. I met her at the entrance of Xingsheng Electronics Factory when she finished work. I saw her riding her red motorcycle heading toward me. She was wearing a pink sweater, which matched her red motorcycle well.

"Wow, you look great!" I greeted her. She was all smiles. She asked me to ride on the back seat and we went to Qingyang Town together.

She had just bought her motorcycle and this was my first ride. On the highway, a lot of motorcycles carrying passengers sped past us and kicked up dust.

"Are you scared?" I asked.

"No, I am not," she affirmed.

As I rode on the back, I found the motorcycle unsteady. So I looked to the front and saw that the handlebar that Ah Zhen was gripping was swaying sideways. I dared not ask anything and held tightly onto the seat. We finally got to the restaurant twenty minutes later. We talked about the motorcycle during our meal. Ah Zhen had spent all her seven years' savings on the motorcycle, which cost her 24,000 yuan. Her father was very unhappy and said she wasted her money. However, Ah Zhen insisted that it was a necessity and she had thought about it very carefully.

She explained her reasoning. "About three years ago, in 1991, I already had the idea of buying a motorcycle. I thought to myself: If I buy it, will I use it very often? What are the pros and cons? The prices are going up; is it worth putting money into the bank? Is it better to buy a motorcycle? Fenkai once said, 'We don't use it very often. Why do you want to buy it?' But I thought it costs me 2 to 3 yuan to ride on a motorcycle from Baixiu Village to Sijing Village, and another 2 to 3 yuan for a bus to Qingyang Town.

Adding up, it costs about 10 yuan to travel to and from Qingyang Town. A motorcycle consumes three tanks of gas each month, which cost 12 yuan each. So it only adds up to 36 yuan a month. Yet, we can have many rides."

Ah Zhen found that the rapid economic development sped up inflation, so she adopted other, more effective ways than traditional saving to keep the value of her wealth. I commended her astute investment and she was proud of her decision.

"What's more, when we grow older, we need 'face.' Frankly speaking, if you put your money in a bank, nobody knows. But if you have a car, even if you are penniless, people will think that you are rich."

She paused and it seemed as if something came up. She changed her tone to one of melancholy: "Perhaps the older I am, the more I need face. I don't want to be looked down upon by others."

So, buying a motorcycle was not just a necessity; it also involved how Ah Zhen saw herself at a deeper level. Possibly it gave her peace of mind.

"While I was riding the motorcycle, the breeze was blowing in my face. It's so pleasant. It's difficult to describe that feeling; anyway it's like all the pressure and worries have gone. I feel free."

"'Free'?" I asked surprised.

"Yes!" Ah Zhen nodded and turned to the topic of the difference of the sexes. "'Freedom' is very important to men. Most women still rely on men and they are obsessed with the idea of getting married and marrying a 'good' man. Some women only nag their husbands for motorcycles, but I used my own money to buy my own bike. It's different when you have to ask somebody for money.

"When I ride my bike, I feel that I'm my own master and it boosts my confidence. Those who depend on their husbands have to look up to their husbands. Easy come, easy go. Besides, only when he gives you more can you spend more. Otherwise, you can't spend whatever you like. It's my money and I can spend it my way. When I'm done, I'll work hard and earn it again," Ah Zhen said proudly.

The "freedom" that she mentioned obviously refers to using her own money the way she liked and deciding for herself how to improve her own life. She does not live off her father, husband, or any other family member. So she does not really care whether her parents agree with her. She only needs to ask herself repeatedly if this financial arrangement suits her. It appears that she begins to understand her own needs based on her personal preferences; at the same time she follows her preferences in her daily life. On that point, Ah Zhen feels that her life is in her own hands.

In the two hours that we were having dinner, Ah Zhen cut me off twice and said, "Wait a second. Let me go and check on my bike." Then she jumped up and headed to the entrance, half-running. A moment later, she came back and, patting her chest, said, "My bike is still there and has not been stolen. I'm relieved!" She smiled uneasily, perhaps sensing that she was over-concerned.

The motorcycle was our topic of discussion at dinner.

"I worked hard for this bike. When I ride it, I am proud. I have a special feeling toward it." Ah Zhen smiled proudly.

9

"A Female's Need for Fulfillment" in the Eyes of Traditional Women

In the previous chapter, we portrayed how Wang Guizhen, a woman who grew up under the reform and opening-up policy in China, understood and interpreted the needs of the self. We described her childhood on the farm and how she benefited from the general education policy implemented under the Four Modernizations when she was a teenager. Through her own efforts, Wang Guizhen graduated from secondary school. Later on she found work in a village–town enterprise. We can see that in both events, when she decided to pursue her secondary education and when she applied for a job, she did a lot of thinking. She considered her own abilities and aspirations. She also calculated the balance between her future income and her daily expenses.

This is a very common phenomenon for young people living in a modern society. Anybody having received basic education will naturally go through such an experience, and this can be seen as part of a person's growth. However, if we put Wang Guizhen back into the setting of a village girl's life in China, we may give some more thought to that "natural" phenomenon. As a result, we might also gain a deeper understanding of Wang Guizhen's growth.

"Follow one's father when at home, follow one's husband when married, follow one's son when old." This motto not only precisely describes the life journey of traditional women in the rural areas; it also clearly states the values they embraced: having a "family" headed by a man is having an anchor in life, a way to achieve peace and stability in life. Establishing and sustaining the relationship between the two sexes also becomes the basis on which they interpret the needs of the self. If we intertwine a person's growth with her interpretation of the self, we can see that throughout the lives of those women living in traditional rural society, familial values direct the entire course of their lives. Apart from the family, they hardly recognize any individual volition. Women participate in farming only to contribute labor to the family, for it can hardly be considered as a productive economic activity for the individual. Women never have any personal wealth outside the family. Of course, we are not saying that under the scope of familial ethical rules there is no personal growth. What we want to point out is that the observations we made of Wang Guizhen regarding her interpretation of the need for self-fulfillment and her growth cannot be solely interpreted on the basis of familial norms. To some extent her story is the result of one's awareness of autonomy being aroused under the reform and opening-up policy. This is something rarely found in more traditional women in the rural areas. If we make a contrast between the lifestyle of more traditional women and the growth process of Wang Guizhen, we gain a deeper understanding of the underlying meaning of her growth experience.

In the first and second sections of this chapter, we will describe the lives of an old and a middle-aged woman. The focus will be on how they view their relationship with their husbands and the next generation. With reference to the change in self-understanding, the third section will conclude the results of the research in Part II of this book.

Li Shimu and Her Offspring

On a winter afternoon in 1994 at around four, I left the village clinic in the market and went to the place where Li Shimu (Teacher Li's wife, as she is respectfully called) lived. This was the second time I had visited her, but I was still not very familiar with the complicated alleys of the neighborhood. My vague memory led me to look for a landmark, which was a bunch of blooming azaleas above the old brick wall. Having made a lot of left and right turns, I still could not find the place. Eventually, I had to seek help from a little girl.

I proceeded according to her instructions. It was raining. The canvas shoes I was wearing became wet with mud, and I felt very uncomfortable. I remembered that last week when I had met Li Shimu, she had spent more time discussing the studies of her grandchildren than the living conditions in Baixiu Village. She had stressed over and over that one had to receive more education in order to get a good job. I indicated that during the weekend when her grandchild came back home from boarding school, I could help him review his homework.

The following week I had been busy visiting other women in the village. Although the elder women I have met since I came to this village were very friendly and warm, they all spoke with a very heavy local accent and they could not express their feelings clearly. Compared to other elderly women in the village, sixty-nine-year-old Li Shimu not only spoke with a pure Guangzhou accent, she was also articulate and could clearly describe to me village life as well as her own feelings about living there. Perhaps this can be attributed to the fact that she had once lived in Guangzhou and that she was also raised in a teacher's family. In view of this, I planned to use her as the focal subject of my interview.

After a few turns, I finally arrived at Li Shimu's home. She was delighted to see me and said, "It's so nice to see you. I was worried that you might not come because of the rain and the cold!" I found that her grandchild Ah Wen had been waiting in the living room. Some paper and books were placed on the dining table.

Ah Wen took out a junior secondary English exercise book and asked me to explain to him what he had done wrong. He was very serious, and he would not let go until he really understood the cause of each mistake. He also told me that he thought the multiple-choice part was the most difficult.

I said, "Perhaps your character is more straightforward and therefore you are not so good at making guesses."

"Right! Not once could I guess anything right."

After a while, Li Shimu brought in a steamed cake. She said this was her grandchild's favorite dessert.

Li Shimu was silent for the next half an hour. She did not watch television, she just kept working in the kitchen. Once in a while she would come by and sit at the other side of the dining table, attentively watching me teach her grandchild. A while later, her eldest grandchild, Ah Hui, was back. He did not make a sound. He sat down in the other corner of the living room and did his own work.

At around half past six in the evening, it was time for dinner. Ruqing, father of Ah Wen, returned home from work. Li Shimu told him that she was lucky to have gotten a tutor for her grandchild. Ruqing then turned to me and gave me his thanks.

Li Shimu brought out a few dishes from the kitchen and invited me to stay for dinner. The five of us had dinner together, watching television at the same time. After a while, Ruqing's wife returned home from work in the factory. She seemed to be about four to five years younger than Ruqing, maybe around thirty-five years old. She said she had already eaten her dinner in the factory, then she settled down in a corner of the living room and started reading the newspaper.

At around seven, a current affairs program was showing on television. The program talked about Hong Kong people buying houses in the Shenzhen area. It said that sellers in Shenzhen promised the buyers that they could transfer the village people's residence registration to the city. As a result, some village people bought properties in Shenzhen through their Hong Kong relatives and turned themselves into city residents. However, sometime this year that policy was changed. A lot of people could not transfer their residence registration. It ended up that they had no entitlement to any kind of residence at all. Among these people, adults were discriminated against in terms of wages and promotion prospects due to their lack of proper residence status. As for children, they could not gain admission to respectable high schools. These people could not return to the village anymore. They were stranded in Shenzhen where the cost of living is high. Life was very difficult for them.

Ruqing said to me, "In the beginning the Shenzhen government wanted to develop Shenzhen, therefore they promoted such a policy. After a few years, the population became saturated and the policy was canceled. Those people who moved to Shenzhen at a later stage ended up 'hanging in mid-air.' Tough luck!"

"I suppose those people would never expect such an outcome," I said.

Mrs. Li echoed my opinion and said, "Right you are! This was especially true a few years ago, when everybody wanted to move to the city, hoping that they would never have to farm again. Who would ever think of a day like today when village people no longer need to farm and still get a share of money every year. Nowadays no villagers want to move to the city anymore. They all like staying in the village."

"Even if they want to return to the village, village people will not want them anymore!" Ruqing added.

Regarding the villagers' preference for staying in the rural area, in fact this is only true for the Pearl River Delta area, where the villagers became rich because of the development of village–town enterprises. For those living in the poor villages in more remote areas of Guangdong Province and for those farmers in Sichuan and Hunan areas, moving to cities or richer villages was still their main avenue to get out of poverty and improve their quality of life. As a matter of fact, Mrs. Li and Ruqing were also aware of this. During the past ten to twenty years, a large group of workers had flooded into this area from other provinces; this was the trend of population flow.

Perhaps they knew that my reason for coming here was to conduct social research. Li Shimu purposefully said more about the situation of Baixiu Village.

"The village has changed. Things that would not happen in the past happen now. A few families nearby have their young daughters in their twenties married to 'Hong Kong folk' almost thirty years older than they are. The sole purpose is to have a better life. To her dismay, the husband of one of those women suddenly passed away, leaving his wife with two kids. The wife is just around thirty. All of a sudden they lost their livelihood. When they got married they should have known their husbands wouldn't have long to live. I can't figure out how they made their plans."

Ruqing did not comment much on these issues. Li Shimu later on changed the subject to the health centers in Baixiu Village.

"The prices here are very expensive, ranging from 10 to 60 yuan. The cost is set by the staff of the village clinic. The same kind of ointment that costs about 6 yuan here only costs about 4 yuan in Xijing Village.

"This is because the income of clinic staff depends on the bonus they can have at the end of the year. If the clinic has a total annual income of 20,000 yuan, the rural government and the clinic will then share a certain percentage of the income with their staff. Therefore, every year when they earn more, the staff will also earn more," said Ruqing.

Li Shimu said, "Among the classmates of Ah Wen in Qingyang Town Secondary School, some come from rich families and have a lot of pocket money. We heard that some of them paid 10 yuan for a piece of bread and didn't even bother to wait for the change."

Ruqing added, "Some of his classmates have 200 yuan pocket money per week, a total of 800 yuan per month. This is a whole month's salary for a lot of people."

Ah Wen added, "Those classmates look so proud. When they have arguments with the teachers, they say, 'My pocket money is even more than your monthly salary!' They just drive the teachers nuts. The teachers in school care a lot about making money. If we have good results, they will get bonus on top of their salary."

When it came to study and future career, Ah Wen told his father, "The most important thing is to earn more money."

From what they said, it seemed that whether it was the registration of residence policy of the Shenzhen government, the move of population from rural areas to the cities, or the day-to-day affairs in Baixiu Village including marriage, medical care, and education, everything centered around the concern for improving people's financial situation. In more extreme cases these concerns manifested themselves as a "money first" social attitude. It seemed that such an attitude had already penetrated all parts of life. People are also inclined to interpret their own needs on such a basis. Perhaps this could partly be attributed to the fast-growing economy. That made the farmers who had had a poor quality of life for a long time suddenly feel the lure of "wealth."

Although Li Shimu did not know how to give a clear analysis of all these issues, she felt something was wrong with all these changes. One example related to her grandchild Ah Wen. Originally Ah Wen was very fond of drawing, and he actually drew quite skillfully. Li Shimu wanted very much to get him to learn art and have him enter the Guangzhou Art Institute. However, Ah Wen himself did not plan to take such a path. Li Shimu told me, "Ah Wen asked me, 'You think there will be a future in art?' He said it in that way. It drove me mad!"

Li Shimu was not really angry; she only felt there was not much she could do, for she knew well what her grandchild said was true. Now, she only hoped that her grandchild would study his secondary school subjects well, especially English, so that in future he could get himself a good job in the competitive job market. This was also the reason why she got me to provide tutorial lessons for Ah Wen.

After dinner, Li Shimu walked me to the gate entrance and asked, "Tomorrow will you come to tutor Ah Wen?"

The next day was Sunday. However, seeing the eagerness on her face, I could hardly say no. I nodded and said yes.

"Then, can you come at half past seven tomorrow morning?"

"That early?" I thought I misunderstood, and therefore I must have appeared a little shocked.

Li Shimu immediately relented, "Then eight o'clock in the morning is fine."

I paused a while, feeling that eight o'clock was still a bit too early. I finally said, "I have something to do in the early morning tomorrow. I can only come at half past eight."

"That's fine. So I'll make breakfast for Ah Wen at eight o'clock tomorrow."

From then on, sometimes I would review Ah Wen's homework on weekends or Sundays. During weekdays, I would chat with Li Shimu. She enjoyed talking about the past. She talked endlessly about her own stories and her family's stories. Sometimes she would share with me her philosophy about life and views about the world. Perhaps this was characteristic of older people, but it also gave me the opportunity to learn more about how she viewed the next generation.

Li Shimu's maiden surname was Shen. Her original homeland was Panyu. Her father had been a teacher in Guangzhou. When she was young, she moved to Guangzhou with her parents. At nineteen she was married to a young teacher surnamed Li, her father's colleague.

"Father purposely introduced to me a colleague of his. He said that he would be more reliable. I married him mainly because he was a nice person. Although he was poor, it did not matter because at that time everybody was poor. No one was rich."

After marriage came the war. Mr. Li went to the villages near Guangzhou to look for a job. Li Shimu followed him wherever he went. The new government came into being in 1949. Mr. Li was sent to Baixiu Village to teach, and Li Shimu joined him. In 1974, Mr. Li passed away, leaving her with four children. Her eldest daughter later married a man in the village surnamed Wang. In 1988, her two younger sons were smuggled into Hong Kong. Those who remained to keep her company were her eldest son, Ruqing, and her daughter-in-law, as well as her two grandchildren.

Li Shimu had been living in Baixiu Village for more than forty years. She knew the village extremely well. Every time she talked about her past, she would first mention that she did not come from a farmer's family.

"I am one who 'never had to carry four taels of rice on my shoulder.' I learned to farm in my thirties. In my whole life, I only farmed during the Cultural Revolution. It lasted for about ten years, during which I spent a few years knitting and sewing. Therefore, my farming days did not last for long.

"The time poverty hit hardest was during the Great Leap Forward period. We heard that because China had to return money to Russia, we all had to get involved in cultivating those 'satellite fields.' We planted several different kinds of crops densely together in the same field. The crops ended up not having enough nutrients. Productivity dropped as a result. At that time we did not have enough food in the village. Life was very difficult."

I asked her how she managed to get through those days.

"At that time I was working as a teacher in a kindergarten. Although there was little food in the village, the adults had to provide decent food for the kids to eat. But adults who worked in the commune only had thin congee."

Li Shimu felt herself lucky: she could be spared from the most hunger-stricken days in the village. The life of a farmer was hard. At times of war or when the government launched inconsistent policies, life could be even harder.

Recalling the family planning policy launched in the village during the 1960s, Li Shimu said, "In fact I had given birth six times. The first birth was a baby boy. I labored for three days and nights in Guangzhou, but still I could not make it. Later on the doctor tried to take him out with a pair of tongs, but they did not do it well. They broke his head and he died. At that time technology was primitive. Even if I had had a smooth delivery, he still might not have been able to survive. The second birth was a baby girl. I managed to have a smooth delivery. Nevertheless, she got a serious flu at ten and eventually died in the Qingyang Town Hospital. The third birth was my eldest daughter Weirong. The following three births were all boys, namely Ruqing, Rucheng, and Ruqiang. After giving birth to Ruqiang, it came the time when the government encouraged us to sterilize.

"It's a painful thing bringing children into the world. After giving birth to them, you still have to bring them up. At that time we were all very poor. How could we raise them? Those who worked on the farm thought the more kids they had the better. That was because they needed people to work on the farm. But I didn't work on the farm. Why did I need that many kids? Some people say, 'Give birth to ten children so that in future they will feed you.' But I say, 'Even if ten of them are all there to feed you, are you going to swallow all that they give into your stomach?' Therefore, when I learned that such sterilization was available, I went for it. There were women who denounced me for being 'too lazy to give birth.' In return I denounced them for being 'close-minded.'"

In fact Li Shimu did not deny the practical advantages of raising children. Now in her old age she had sons to support her. She did not have to worry about food and shelter. Occasionally she could travel abroad. In a way, she did experience the effectiveness of "raising a son to prepare for old age."

The two younger sons of Li Shimu, Rucheng and Ruqiang, were smuggled into Hong Kong in 1988. Both of them had gotten married and were raising their own children. They lived in public housing provided by the Hong Kong government. Ruqing had visited his two brothers with his mother. I heard from them that their brothers led quite a good life. They had very modern household appliances, although the units they lived in were rather small. The first time I had dinner with Ruqing, he said to me, "In Hong Kong you have everything. You eat well, dress well. My younger brother's daughter is learning English at a much more advanced level than Ah Wen and Ah Hui. You people also earn much more money than we do."

In fact, in 1974, Ruqing did try to sneak into Hong Kong. He made his attempts even earlier than his brothers did, but he did not succeed. Li Shimu told me, "At that time his father had just passed away about a hundred days earlier. It was close to Mid-Autumn Festival. It was raining hard. Ruqing and four of his friends hoped the guards at the border would be more relaxed. They took a few moon cakes and some bottles of water with them and left. After the young men had walked a few days and nights, the cakes became drenched by the rain. They became mashed and were too repulsive to eat. One of them could not stand it anymore and returned home. Ruqing and the other three continued on their journey. When they finally came upon a village, they were on the brink of exhaustion. The people in the village asked them what they were doing. Because of their exhaustion, they lost their good sense and blurted out 'We are fleeing!' They were detained immediately. Later on Ruqing escaped, and he went on by himself. Finally he reached Shenzhen. On the mountain he could see Yuen Long in Kowloon. However, at this point he was captured by the police."

Since then, Ruqing did not try to flee anymore. He stayed in the village. Later on he got married and raised his own family.

Nonetheless, seeing that his two brothers succeeded in escaping in 1988 on their first attempt and seeing the kind of life they had in Hong Kong, occasionally a slip of the tongue from Ruqing would tell that he was not very happy. It seemed that when compared to his brothers, Ruqing was not doing that well. However, as far as Li Shimu was concerned, deep in her heart she had not wanted to let her sons go. She was also concerned about their safety during the journey.

"A lot of people did not return. They either drowned or fell to their death in the mountains. Despite that, the hard fact was, we had no way out. It was too poor in the village. Furthermore, nothing could stop them if they wanted to leave," said Li Shimu regretfully. Although poor village life was their reason for fleeing, and it was a good reason, Li Shimu still felt bad about her son wanting to leave right after her husband died. This to some extent indicated that her son regarded his father as more important than she. Nevertheless, living in a culture that traditionally look to men for the lead, it seemed that a mother could only accept the reality and adapt herself to it.

For many years in the past, what always had been in Li Shimu's mind was how to finish the unfinished duties of her husband—that is, how to help the next generation establish their own families.

"When looking back, I would think if all three sons were here, I would have a hard time trying to build a house for each of them. It took 60,000 to 70,000 yuan to build a house even five years ago. Where could I get that much money? And how would I be able to look after all their kids? It's fortunate that we only have Ruqing here. At the time when Rucheng and Ruqiang fled to Hong Kong, I never thought of this. Maybe that is how all things in life would turn out—unexpectedly.

"Weirong is a person who 'would grab a handful of sand from the ground in case she falls onto the ground': she just never admits any fault on her part. That daughter of mine is efficient and diligent, but she always picks fights with me. Therefore, I deliberately arranged a marriage for her in which she did not have a mother-in-law."

Deep in Li Shimu's heart, what mattered most now was to improve her grandchild's academic results. She would play mahjong with other women in the village when she had some free time; otherwise, she would take care of the family chores. One time, she chatted with me as she was cooking vegetables in the kitchen. She said, "There are not many days left for me now. Nevertheless, all my children have grown up, gotten married, and reared their own kids. I told them, 'It's fine with me to die at any time. You don't have to cry over my death.'"

Dayan Sao

Dayan Sao (Aunt Big-Eye) was a fat woman of around thirty-five. Like other women her age, her school years happened to fall within the Cultural Revolution. Her family did not encourage her to go to school. Dayan Sao stopped attending school after grade five. In her early years, farming was her livelihood. Later on, enterprises boomed in the towns and villages under the reform and opening-up policy. During that time she worked at a factory. Eight years ago, when she was twenty-seven, a matchmaker arranged a marriage for her with a Wang in Baixiu Village. Before her marriage, she lived in Xijing Village nearby. Her husband, Mr. Wang, was ten years older than she. In his early years Mr. Wang was smuggled into Hong Kong to make a living. He came back four or five times every year during festive holidays to visit his relatives in the village. After his marriage, Mr. Wang continued to live in Hong Kong. Dayan Sao stayed in the village and lived with her mother-in-law, a son, and a daughter.

Dayan Sao spent every day the same way. She made breakfast for her seven-year-old son Yaozu and three-year-old daughter Mingming at seven o'clock in the morning. Then she sent them to primary school and a childcare center. At eight she rode her bicycle to Xijing Village to buy groceries as well as to drop by her mother's home. She would then chat with her mother and younger sister. At around ten she rode her bicycle back to Baixiu Village. When she got back home she did her housework and prepared the meals. When Yaozu and Mingming came home from school in the afternoon, she would bathe them. After dinner, she would watch television programs from Hong Kong, then she would go to bed. The next day she would rise early in the morning and repeat the pattern.

Dayan Sao usually squatted beside a well in the front yard to wash clothes. Every time I came to visit her, she would bring me a small wooden stool and ask me to sit down and chat with her.

When we discussed her lifestyle, she said, "Life now is no doubt more comfortable than the time before I got married. Now my husband sends money home every month. However, it is very boring to stay at home all the time! I have mentioned to him that I want to go out to work, but my husband does not want me to do so. He wants me to take care of the kids. I have also mentioned to my mother-in-law that I would give her 100 yuan to look after my kids. She refused me. I always have fights with my mother-in-law. She is always rattling on. She is so long-winded that she makes me feel irritated and agitated."

Dayan Sao's mother-in-law, Wang Popo (Old Woman Wang), was more than seventy years old. She was also married into Baixiu Village from Xijing Village. Usually she enjoyed playing mahjong with other old women in the village. Recently she did not play very much. Li Shimu told me that Wang Popo had some conflicts over money with the people she played mahjong with, and therefore recently she spent more time at home.

In my opinion, Wang Popo was not as warm as other old people when receiving guests. Nevertheless, every time I visited them, she would chat with me for a while. She spoke with a heavy village accent. She always complained that the law and order in the village was getting worse and that the social attitude was getting poorer and poorer. She also complained that the next generation was no longer "diligent in work and obedient to parents." Alone, Wang Popo always stayed in the living room and burned incense to worship the ancestors of her husband's family. She was serious and respectful. Dayan Sao would only do so during festivals; even then her heart was not in it. It looked like she was just half-hearted in finishing some duties. In fact, other middle-aged women in the village held the same kind of attitude toward worshipping ancestors, only some of them did not display this so candidly in front of the elders or their husbands.

Although Dayan Sao and her mother-in-law lived in the same house, they cooked in their own kitchens. Dayan Sao's kitchen was in the front yard. It had everything including a stove, propane gas container, and washing machine. Wang Popo's kitchen together with her bedroom was on the left side of the living room, always locked up behind an iron gate. Looking into it from outside, one could see only darkness. It was obvious that the relationship between the two in-laws was bad. Once I saw them have a fight; it was triggered by a conflict over choice of television programs.

Wang Popo enjoyed watching a Hong Kong television program called "Zhen Zhu Qi" (The Pearl Flag), filmed in an ancient setting. Dayan Sao liked to watch a contemporary program named "When We Meet Again You Are Still My Wife." This program was broadcast by another television station in the same time slot as "Zhen Zhu Qi." The contemporary program was about a diligent and promising merchant who had a love affair outside marriage. In the end, his mistress forced his wife to leave him as well as her children. Later on, the merchant's business failed, and the mistress left him. He regretted what he had done and asked his ex-wife to come back to him. However, by that time, his wife already had a new boyfriend. The other program, which was an old classical tale, was about the seven widows of the Yang family in the Song Dynasty. They waged war with alien tribes and fought on behalf of their husbands, who had died in war. One of those seven widows was young and a good fighter. She fell in love with a young knight. Under the guidance of her mother-in-law, this widow finally refused the love of this knight. The story reflected her unfailing love for her late husband.

Wang Popo said, "Of course 'Zhen Zhu Qi' is more entertaining. What good is it for my daughter-in-law to watch those programs that talk about a husband going after other women and a wife going after other men? Women in the past were all gentle and virtuous, not like the women now. See how nice the widows of the Yang family are treating their mother-in-law and their deceased husbands! They are not like today's women; some of them even ask for divorce."

Dayan Sao looked displeased. She said, "If a woman cannot stand her husband having another woman outside marriage, she just cannot! Frankly, nowadays people choose their own husbands. If it is not something really unbearable, who will go for divorce?"

Wang Popo said, "What's so hard for the women to take nowadays? It was even worse for us in our day. A lot of husbands in our generation took opium; they also gambled and went whoring. Can you say those women of the past did not suffer more than you women of today? In the past there was a popular story going around. It was about a woman deserted by her husband—he kicked her out of the house. Having walked a few miles, the woman carelessly lost her hair pin in the snow. While she was searching for her pin, an elder passed by. The elder asked her what she was doing, and she said that she was looking for her pin. The elder recited a poem that goes, 'Who pities an old hair pin. . . .' Having heard that poem, the woman decided to return to her husband."

Wang Popo explained to me the moral of this story with a village accent that got ever thicker. The gist of it is that although the hair pin was already old and broken, the woman was still unwilling to throw it away. That was because she had some indescribable feeling toward this pin that had accompanied her for many years. Wang Popo repeated the words of the poem and kept murmuring, "No matter what, one still should not leave one's husband. No, never!"

"You are so outdated. Times are different now." Dayan Sao threw her mother-in-law a glance, then turned to me and said, "What's the big deal about husband and wife? In the last analysis, a husband is not reliable. Nowadays women have to rely on themselves and their kids. Your kids are reliable because you gave birth to them. A husband marries you, takes you home, and it is like acquiring an item of clothing. He does not treasure you anymore after marriage. Times are different now. If a woman cannot take it, it's no big deal to get a divorce!"

Having heard these complacent words from her daughter-in-law, Wang Popo retorted immediately, "How can one do that?"

"Don't bother with her. She is not a reasonable person!" Dayan Sao said angrily.

To calm the situation, I changed the subject. I asked Dayan Sao how a divorced woman lived her life and what kind of attitude people in the village held toward them. She had not even uttered a few sentences when Wang Popo interjected, "They have to marry again as soon as possible, as soon as possible! When one gets old it's hard to find a man!"

"How can you just get divorced and then get another man to marry again right away? Even if you are shopping for groceries, you still need time to select what you want, not to mention looking for a husband!" Dayan Sao retorted.

Wang Popo paid no attention to what her daughter-in-law said. She just kept on mumbling to herself. It seemed as if she was advising divorced women to get remarried as soon as possible. Dayan Sao appeared to get flustered, but did not say anything anymore. From time to time she would look at her mother-in-law indignantly.

The squabble between the two in-laws came to a pause here. They often fought over large and small issues in the household. Dayan Sao had gone through primary education. This, coupled with her character, meant she always got the upper hand in such petty fights. Wang Popo, who did not know how to read, was always at a disadvantage. Despite that, Dayan Sao dared not go too far. There was something restraining her.

Once she mentioned to me, "There were rumors going around that my mother-in-law went back to Xijing Village and complained to my mom. She said that I always scolded her. My mother therefore scolded me. She wanted me to give in more as a daughter-in-law. But honestly, I want to express myself. One has to voice what one does not feel happy about! How can we be like them—the older generation?" She paused a while, as if searching for something. Then she said, "Well, forget it! Let her watch 'Zhen Zhu Qi.' There is no point quarreling with her anymore."

The relationship between mother- and daughter-in-law has always been a very complicated matter in Chinese families. This is not only an issue of the generation gap. It is related to a familial system that makes the male the center of power. It makes the two women, who live under the same roof, wrestle with each other for power and living space. In a culture that stresses "order according to seniority" and "men superior, women inferior," a woman has to move from the position of daughter-in-law to the position of a mother-in-law before she can attain a raise in terms of power and position and earn the respect of the son. A daughter-in-law has to obey her mother-in-law: that is what was demanded of daughters-in-law during the time of Wang Popo's upbringing. However, Dayan Sao was brought up under the Communist regime. It was difficult for her to accept wholesale the set of rules that came down from tradition. From the argument I had just witnessed, one could see that Dayan Sao disagreed with her mother-in-law mainly because Wang Popo only aimed at sustaining a marriage and paid no regard to whether the relationship was built on the feelings a couple had for each other. For the women in the older generation, it was natural to depend on marriage as a source of livelihood. In fact, even Dayan Sao did not deny this, but she also thought that the inner feelings of a woman were very important. This is where their differences occur. Nevertheless, if we observe more carefully, we see that what Dayan Sao said about divorce was not really how she viewed it. She totally refuted Wang Popo's views because she was having a fight with her. Later on I had a chance to discuss that television program with her again. Then she told me what she had really meant.

"If I were the wife in the story, I would choose to reunite with my former husband. Although he once had a woman outside marriage and he had once acted callously, I would still reunite with him because of the kids." She paused a while and said, "Furthermore, the elder daughter she had with her former husband was already about ten years old. She did not have to start all over again if she went back to her husband. This way she could save ten more years' time. Although she could still have children with the other man if she got married to him, she would have to start all over again."

So it seemed that in fact Dayan Sao was even more astute than her mother-in-law in planning for her own livelihood. The only difference was that Dayan Sao focused more on her children as her future means of support. In fact, during the last argument, she also mentioned that a husband's feelings toward his wife would change after marriage sooner or later. Perhaps this was her personal experience from real married life.

"We as women will get old, no matter how beautiful you were when you got married. It is likely that a husband will seek other women at such a stage. But then, it would be difficult for a woman to get another man at such a time."

Having been married for eight years, Mr. Wang lived in Hong Kong most of the time. He only came back to the village for family reunions four or five times a year during longer holidays. As such, Dayan Sao spent the whole year waiting for him in the village.

Once, around the Lunar New Year, I said to Dayan Sao, "Only two more weeks to go and Mr. Wang will be back. You must be very happy!"

"Right! But I have to wait so long before he's back, and every time he only stays a few days and then has to leave again. I really don't want him to go, but there is no other way."

"Though it is a short stay, you will have a few days together after all. It's still something to feel happy about," I comforted her.

"You are right in saying that. However, my husband cannot come home during other times because he has to work in Hong Kong. He can only be back during festive seasons such as Lunar New Year and Mid-Autumn Festival. When he finally has the time to come home during festive holidays, these are the days when I become busiest. I have to worship the gods, make meals, and so on. When the festival is over and I can have more time, it's time for him to return to Hong Kong. After he leaves I have to stay home and feel bored again. It's like that every time." Dayan Sao looked very helpless.

A lot of women in the village who were married to Hong Kong people faced a similar situation. Before they got married, they knew that husband and wife would stay apart more than together after marriage. However, the income earned in Hong Kong was far more than that in the village. Life would be better this way. In the last ten years, a lot of older women who could not find a husband in the village got married to village men who had left for Hong Kong in earlier years. Every year near festive seasons, when I chatted with these women, they all said that they were eagerly expecting their husbands to come home for a reunion. I thought no matter how short these festive holidays were, they would definitely bring some joy to Dayan Sao. Nevertheless, things did not seem to be going as well as I thought.

She said to me, "During the few days when my husband stays home, he always likes to play mahjong. I don't like him to gamble, but he says he has gotten used to it and he cannot quit a habit that he has had for that many years. There was one time he went gambling and I waited until midnight. Still he did not come home. I was so angry that I locked the main door and refused to let him in. I left him yelling outside. I did not care. You know, I can be like that. When I become fierce, I can be very fierce. My mother-in-law can do nothing about it.

"I said to my husband, 'You are not young anymore. If you don't save up some money now, when you are already close to fifty, what will we do when we get old and our son is still too young to make money to support us?' But he never listened to me. Every time he lost all the money he earned.

"I've told him the same thing many times and I don't have the energy to repeat it anymore. As long as he gives me money for the family every month, I leave him alone." Dayan Sao paused a while and said, "But he has to give me money. There's no way he does not give me money. If he gives me no money I'll be fierce. He should know I am very serious about money."

So the fact is, even if a woman succeeds in marrying, she still has to worry about her livelihood. Obviously, this was not something Dayan Sao could have predicted before she got married. Nevertheless, like other women in the village, she considered marriage the most important issue in a woman's life. Originally I thought this was only a matter of "having an anchor in life," but later on I found that at the back of her mind there were other calculations.

"One gets tied up after marriage, but a woman has to get married anyway, and preferably earlier. There's no doubt that one feels freer before marriage. But then even if you have one or two more years to make money, there still will come the day when you have spent all your money. What can you do then?

"If one gets too old, for example if one gets married after thirty, she will have much pain giving birth. She'll have to undergo an operation to deliver the baby, and that will be very painful. The technology here is very primitive; wounds take a very long time to heal."

From the topic of getting married and giving birth, we went on to the topic of raising children. "And there is also the issue of rearing the kids. If a woman doesn't give birth until she gets old, then when she gets really old, like at 50, her kid is only 10 or so years old and still needs to be taken care of. When you look at yourself in the mirror at that time, you'll say to yourself, 'You've become old!' This way you not only make yourself suffer, you make the kids suffer too!

"If a woman gets married earlier, even if the husband visits prostitutes, or even if she get abused or deserted when she's old and nobody wants her anymore, still she can rely on her son! That's why I say, 'A husband is dear, but a son is dearer.' The most important thing of all is to give birth to a son."

Dayan Sao considered her own thinking very insightful. As a matter of fact, practically speaking, she had thought thoroughly over the pros and cons of getting married and how she should plan for her life after marriage. There is one point we should pay close attention to. Presumably, a husband should have an important position in a marriage, but what Dayan Sao said mainly centered around how a woman's age at marriage would affect her age when giving birth, and how her age when giving birth would affect the age difference between the mother and the son, and how this difference in the end would affect the way they took care of each other. In the eyes of Dayan Sao, the son seemed to have replaced the husband and become the only person a married woman could trust and rely on.

She took the teaching of her son Yaozu very seriously, and she explained to me the importance of educating and disciplining one's children. "Under the policy of family planning, usually people in the village are only allowed to give birth to one or two babies. The kids therefore all get spoiled. If people are not strict in teaching their own kids, in the end they themselves will suffer."

She said that she always tried her best to grasp any opportunity available to teach her son the principles one should uphold in life.

"When I come across some bad characters on the television program, I'll remind Yaozu, 'Don't do what these people do. If you do so and turn bad, you'll become a useless person!' I was very serious when I said so."

Discipline was a principle Dayan Sao insisted on teaching her son. The pomegranate tree in the front yard was the "teach-the-son" tree where she gave Yaozu lessons.

Once Yaozu played outside with other children in the village and his pair of shoes was stolen. He had to walk home barefoot. Dayan Sao asked him what happened; Yaozu dared not say a word. Dayan Sao raised her voice and asked again. Yaozu was stunned and became too frightened to utter a word. Dayan Sao was angry and so she tied him up, hung him up on the tree, and gave him a good spanking.

Yaozu was a big, fat boy. He had an "abalone brush" kind of haircut. As Dayan Sao did not regulate his diet when he was small, now at seven years old Yaozu's waist already exceeded 75 centimeters. Everybody in the village called him "fat boy."

When I heard that Yaozu was hung on the tree and beaten, I was stunned. "Oh dear! He is so fat, wouldn't it be very painful for him to have both hands tied and hung onto the tree? Is it really necessary to punish a kid like that?" I asked.

It seemed that Dayan Sao did not hear what I said. She rambled on in higher spirits, "I had to punish him. My mother-in-law was standing beside me and she pleaded with me on his behalf. She said, 'Don't do that!' I yelled at her, 'Don't you ever untie him!' You can tell I'm very strict with my kids!"

Another day, I purposefully mentioned this "teach-the-son" tree incident again. Dayan Sao did not find anything wrong with such a teaching method. She stressed that when she had tied her son to the tree, his feet could still reach the ground.

Just then, Yaozu, who was playing with the bicycle under the tree, pouted and said, "It didn't hurt being tied to the tree, but it sure hurt when she beat me!"

Dayan Sao said if her little daughter became naughty, she would treat her just the same. She was very confident. She felt that she had more foresight than other mothers because she would not spoil her children. In Baixiu Village, women who have to go to work usually do not have time to teach their children. Very few housewives would take seriously the issue of teaching and disciplining their children. Compared to other middle-aged women, Dayan Sao no doubt gave much more serious thought to this issue. Nevertheless, seeing the way she taught her children, it made me wonder if she had not transferred all the frustrations of her day-to-day life onto her kids.

Once, when I was chatting with her under the "teach-the-son" tree, Dayan Sao seemed unable to suppress the frustration deep down in her heart, saying, "It's very tough to rear the kids. If my husband lets me, I'll surely go out to work. This way at least I can talk to people and won't feel so bored. Taking care of the kids is a tough job and I always feel agitated!"

Conclusion: The Change in Self-Understanding

In the first and second sections of this chapter, we portrayed the lives of Li Shimu, Wang Popo, and Dayan Sao. We see how these more traditional women in the village use familial values to interpret the needs of a female, in contrast to the young woman, Wang Guizhen, of the previous chapter. In Wang Guizhen's case, we saw how she had gradually used her personal aspirations and interests as the basis on which she interpreted the needs of her self in a social environment under the rural economic reform. From the point of view of our research, we see that Wang Guizhen's growth to some extent manifests the rise of an awareness of autonomy under the reform and opening-up policy. It also manifests some changes in the self-concept of village women of a younger generation.

We shall briefly revisit the three stories, then select some main points for further discussion and outline what kinds of changes have taken place with regard to village women's self-understanding in the young, middle, and older generations. Hopefully this will help us to capture the changes in the development of Chinese people's self concept from traditional to modern times.

In this research, we make many efforts to interpret the subjects in the specific social and cultural contexts in which they belong. The social changes brought about by the reform and opening-up policy in 1979 are the basis of this research. Nevertheless, we are not taking this period out of the context of China's recent historical development. Instead, we are comparing the social conditions of this period to those of the two earlier periods, namely the period since the inception of the People's Republic of China in 1949 to the time when the Cultural Revolution ended, and also the period before the inception of the People's Republic of China. As for the earlier period, we notice that after the

Chinese Communists came to power, they immediately launched a series of policies under the slogan of "antifeudalism" in order to deal a heavy blow to lineage activities. These included the destruction of ancestral halls, forbidding marital ceremonies related to lineage, encouraging couples to move out of their parents' home or to live on a separate floor if they continued to live together. All these policies challenged the familial concept of the traditional village. Dayan Sao was born after the beginning of the People's Republic of China, and she grew up during the period of the Cultural Revolution. She was influenced precisely by these policies. We will discuss her story later on. Now let us look at the situation of the elder women from the generation before that of the People's Republic of China.

In the years before 1949, the traditional cultural beliefs and value system that governed the daily life of Chinese people were still deep rooted despite the fact that China had been ravaged by years of war. "Follow one's father when young; follow one's husband when married; follow one's son when old" is an exact description of Li Shimu's life. When she was young, she moved with her father to Guangzhou City. Her father later on arranged her marriage for her. After marriage she followed her husband and moved around in the village areas near Guangzhou City, eventually settling down in the Baixiu Village of Qingyang Town. After the death of her husband, she lived with her eldest son, taking care of her grandchildren. Although Li Shimu knew her son would not treat her as reverently as he had treated his father, what she was concerned about throughout this ten-to-twenty year period was how to arrange for the marriage of the next generation and how to help them establish a family. What she did for her offspring was obviously a way of finishing the duties her deceased husband could not finish. This was the goal of Li Shimu's life. It was also the basis on which she interpreted the duties of her self.

Alone, Wang Popo often burned incense to worship her husband's ancestors. Her respectful and serious attitude was rare among the middle-aged women. In the scene where the two in-laws quarreled, the daughter-in-law was completely against the mother-in-law. In the end, she got the upper hand over the old woman. Nevertheless, if we view it from another angle, we see that Wang Popo did say from her heart what she truly felt despite the fact that she was not as articulate and argumentative as her daughter-in-law was. Although there were times when she appeared to be contradicting herself, this only reflects the fact that there could be problems embedded in the traditional values themselves. No matter what, the more Wang Popo desperately insisted on the values that she could not explain well, the clearer it demonstrated the commitment she had in sustaining a life-long marital relationship.

From this, and also from the life lived out by Li Shimu, we can see clearly how traditional village women interpreted and understood the needs of the self. A woman had to be somebody's wife, produce offspring, and follow the familial practices of the husband's family. Only by doing so could she help continue her husband's lineage, partake in the customs and practices of her husband's familial society, and be identified with her husband's family and lineage. It was through such a moral link that a woman could obtain peace at heart. This existential feeling forms part of the "ultimate concern" that was discussed in the first chapter of this book. We can also take this as the basis on which to discuss the relationship between the two sexes.

Nevertheless, as we have pointed out in earlier chapters of Part II as well as in Part I, traditional marriage in China, including the marriage between Wang Popo and her husband, is to some extent function-oriented. A man covets a woman's ability in providing labor and producing offspring. On the other hand, a woman relies on a man to provide her with food and shelter. We want to point out that such a phenomenon is brought about by the mentality of traditional Chinese. An individual always has to transform his or her personal aspirations or even self-interests into the greater meaning of fulfilling the grand missions of the family. Only in such a way can an individual get recognition from the society and his or her actions become justified.

Let us see how this mentality is applied to a woman. Wang Popo, a member of the traditional generation, dared not make it so plain as to say the purpose of getting married was to have a husband to rely on. She would stress more the moral importance of "having an anchor." Under the grand principle that one should act for the good of the lineage, she would regard as justified the personal sacrifice of a woman who tolerated the unfaithfulness of her husband. She even used expressions of praise such as "gentle and virtuous," "faithful to the husband," and "being loyal to only one man throughout one's life" to show her admiration of these women. Nonetheless, while we analyze it this way, we are not saying women of this generation deliberately, consciously, and strategically packaged their personal interests in the form of moral values. Quite the opposite, what we want to say is that living in a traditional Chinese culture where there was no rationale for the independent existence of individual aspiration, feelings, and interests, Wang Popo could only attach her own personal interests to the moral values of the lineage. That was the only possibility, and the only way available to her. At the same time, she would have to take these moral values of the larger self as the basis on which she actualized the needs of the self. Although in some undefined ways, individual interests and familial order were not compatible, there was no other choice for Wang Popo, who lived in the homogeneous value system of the traditional familial society. Every time she came across an unhappy experience, the only way out for her was to resort to lineage moral values. In doing so, she could transcend the pain behind her consciousness as an individual.

However, when it comes to middle-aged women who grew up under the slogans of "antifeudalism," the authority of taking the familial ethical order as the basis for interpreting the meaning of individual existence was greatly challenged. There were also signs that such authority was gradually being abandoned. The conflicts between Dayan Sao and her mother-in-law obviously reflected the fact that in Dayan Sao's eyes, the family ethical order was no longer something too holy to be challenged, especially when it came to the relationship between husband and wife. The moral value behind this relationship of the two sexes no longer solely rested on familial ethical order. Instead, the micro aspect of an individual's feelings also had to be considered. When a woman's feelings were hurt because she found out that her husband was unfaithful, her feeling of pain could become the rationale behind a decision of whether the relationship should go on. This can be seen as an important characteristic of the modern marriage concept. Nevertheless, as mentioned earlier, in the traditional thinking of Wang Popo, an individual's behavior had to be justified by the moral values of a familial society, not by the feelings of the individual. Therefore, she would criticize a wife for disregarding the interests of the family to follow her own will and leave her husband. In the eyes of Wang Popo, any individual act that challenges the familial ethical order was viewed as an immoral selfish act. What is worse for a woman was that once she had disassociated herself from her husband's family, she would immediately lose the support that would give her peace at heart. She might even be unable to secure food and shelter for the rest of her life.

Obviously, Wang Popo's interpretation, expectation, and value judgment of the relationship between the two sexes all originated from the times in which she was brought up. Like Li Shimu, she is a product of the traditional Chinese generation before 1949. Expecting daughters-in-law to be "gentle and virtuous" and "loyal to only one man throughout one's life" were traditional moral standards that Wang Popo insisted on from the beginning to the end. When she was young, she did go through such a baptism by fire. However, times changed. Dayan Sao was born after the People's Republic China and she grew up during the Cultural Revolution. Although she did not go to the extent of negating the entire set of traditional values, it would be difficult for her to follow the old set of standards willingly and wholeheartedly. As a result, there is conflict between the two in-laws in this section. This common generational problem is not solely caused by age differences. Nor is it a problem simply caused by the unequal structure in a family that creates the conflicting roles of mother-in-law and daughter-in-law engaged in a kind of political tug-of-war under the same roof. What is more important is that it signifies the existence of conflicts between two different value systems in the wake of political change in China. Accompanying such a change, political awareness penetrated rural society, and economic reform also influenced people's thinking. As a result, a new value system was created, conflicting with the old. The mother-in-law believed that the familial values were enough to override any feelings of an individual woman, but Dayan Sao to a rather great extent wanted to give recognition to the independence of the individual female feelings.

As a matter of fact, Dayan Sao expressed support for divorced women in the village on the basis that she recognized the individual feelings of a woman. She also relied on her feelings in deciding to marry to her present husband while she still had other choices. Our research shows that this reflects the budding of a certain kind of awareness of individualism. Dayan Sao had a better understanding of her own feelings; she also stressed the importance of individual choice. This exemplifies what we discussed in chapter 1 about the development of "autonomy." Nevertheless, we need to qualify this. From what we could see, Dayan Sao had not considered her marriage a good one. For an explanation, we have to examine the social environment that village women are currently in.

In the previous chapters it has been shown how under the reform and opening-up policy, Baixiu Village saw the emergence of a social phenomenon in which people based everything on "economic development." Commercial and industrial productivity have replaced farming in recent decades. Village people learned what constituted rational calculations in trading. They also learned how most effectively to satisfy their self-interests. This rational outlook made individuals start to value and recognize their own personal interests, especially their material needs. They also looked upon all the people and things in the outside world as tools to satisfy their personal desires. As a result, relations between people became mainly function-oriented. We can say that the wildfire spread of prostitution and of "keeping a mistress" in Baixiu Village reflects clearly how this ideology has distorted relations between the sexes. This ideology challenged Dayan Sao's confidence in her marital relationship. She stated firmly that she could not put her trust in her husband; her livelihood ultimately relied on "the son." Furthermore, she used a set of relatively sophisticated "mathematical formulas" to explain the traditional concept of "rearing a son to prepare for old age." From this she further deduced how marrying later in life was unfavorable for a woman. We can see clearly that when Dayan Sao could not manage the emotional relationship with her husband, she immediately expressed anxiety about food and shelter needed for the rest of her life. She was also influenced by the concept of a function-oriented relationship between the sexes; therefore, she viewed the relationship between herself and her son from a functional angle as well. The fact that Dayan Sao took such an attitude is possibly partly due to her lack of financial independence.

As for members of the younger generation like Wang Guizhen, the reform and opening-up policy facilitated the development of independence in the individual both financially and in other aspects. In chapter 8 we described how Wang Guizhen had experienced a tough but enjoyable childhood on the farm. We also describe how she benefited from the general education policy launched during the Four Modernizations period after the Cultural Revolution. This enabled her to finally obtain a complete education, unlike village women of the previous generations. Having a secondary school education, Wang Guizhen became one of the few elites in Baixiu Village. Although people in the village did not give her due recognition because she is a woman, Wang Guizhen did successfully obtain a job at the middle-management level of the village–town enterprises because of her relatively high education level and capability.

We do not underestimate how the "male superior, female inferior" concept in the familial culture has retarded Wang Guizhen's development. Nonetheless, generally speaking, she still has plenty of potential for development, more than the older-generation village women, particularly with respect to individual consciousness. In contrast to Wang Popo, the metamorphosis in Dayan Sao is possibly her verbal expression of personal feelings. In Wang Guizhen's case, the change is in the fact that she is even more articulate in expressing the joy, sadness, fear, and even anger. When faced with the problem of marriage, she took as the starting point her personal feelings and raised the question of whether the two sexes could build a marriage without some kind of communication as a base. That is a question hardly heard from more traditional village women. Even if they did have such a concern, they would not act like Wang Guizhen and query the reasonableness of an existing system simply because of their own personal feelings.[1] Furthermore, Wang Guizhen, who had received a higher level of education, also knew better than Dayan Sao how to use her rationality to analyze her livelihood as well as to plan her own future. We shall not elaborate this point further. Nevertheless, what is worth mentioning is the expression "freedom."

At the end of chapter 8, Wang Guizhen used the term "free" to describe the feeling of driving a motorcycle. For people living in modern Western societies, "freedom" is definitely not a strange word. It is not only an abstract academic concept for more in-depth study, but also a value conviction rooted in the daily life of modern people. It is not only an indispensable part of the modern self-concept, but an indispensable guiding principle in establishing a modern social system. As there is such an inseparable relationship between "freedom" and "modernity," it is even more worthwhile for us to contemplate how Chinese village women in the reform and opening-up era who had lived under the Communist regime expressed their concept of "freedom."

Li Shimu and Wang Popo never mentioned the word "freedom" to me. They silently fulfilled the responsibility of doing the best for their husbands' families. They did their best to be "gentle and virtuous" women. From this angle, the concept of "freedom" did not seem to exist in the world of the older-generation women. Middle-aged Dayan Sao did mention that before marriage one had more freedom, that after marriage one was confined to the home and had to repeatedly do the never-ending housework as well as take care of the children. Here Dayan Sao did reveal her desire to break away from the bondage that restricted an individual's actions. When it came to young Wang Guizhen, she pointed out specifically that "freedom" meant a woman did not have to rely on a man for financial support, and that she was able to use the wealth she earned from her labor to determine the way she lived.

If we take Wang Guizhen's interpretation of freedom as the preliminary model of modern autonomy in China, we might be able to further trace the social conditions that facilitated the formation of such autonomy. Wang Guizhen's words indicated that "financial independence" was the basis for female freedom. Let us analyze her views on the dual levels of the individual and the social structure.

The former means that an individual possesses private property. If an individual's wealth comes from productive activities outside her own family system, and is not supplied by the family, then the individual has the absolute power to decide how to use her wealth. Even parents and husband cannot trespass on her rights to such wealth without the consent of the individual. In such a situation, an individual needs to think independently and make a decision as to how to use her personal wealth. Wang Guizhen is a typical example in that she bought her motorcycle despite the objection of her father. We do not have to consider whether it was right for her to do so or not. What is worth noting is that the possession of private property has enabled an individual to break away from the expectations of lineage and family when she exercises her rights. In this way she frees her personal desires from the bondage of familial expectations.

There is no doubt that "private property" is an important concept in nurturing the modern concept of autonomy. However, that needs to be accompanied by a corresponding development of social structure before it can come into existence. In a traditional Chinese rural community, the organization of economic productive activities was based on the network of family relationships. The familial relationship network itself is a normative order, in other words, a model of moral life. As a result, economic activities and normative order merged and became one. There was no clear boundary between the two. If familial norms could control various areas of social life and maintain a homogeneous value system, then an individual could only rely on this one source to interpret the meaning of her own existence. She could only interpret the needs of her self on the basis of the moral standard of the larger self, which was the traditional way of self-understanding. However, when the small-scale farming economy operated by familial societies was replaced by incoming large-scale industrial production, economic life was gradually uncoupled from the familial structure and formed an independent system of its own. This "uncoupling" was not merely a process of disintegration of social structure; what is more important is that it facilitated the disintegration of the homogeneous value system. Those areas of social life that had become independent would develop their own value systems. If an individual worked for an industrial production facility outside her own familial structure, then there was a possibility that she could develop her own set of values while she carried out her own independent economic activities.

In chapter 8, Wang Guizhen, who worked in a village–town enterprise, is a typical example of how a younger-generation village woman can develop individual autonomy while engaged in the economic productive activities of industrialization. When contrasted with rural women of the older generation who in various degrees had abided by the traditional concepts of life, Wang Guizhen's growth and her interpretation of her own needs seem to show that a process of change is taking place. That process is one in which the basis for a woman's interpretation of her own needs has gradually changed. It represents a turning away from the familial ethical order and moving toward the individual's personal feelings, to contribute to the budding of individualism.

Part III

Sex and the Sex Trade Under the Reform and Opening-Up Policy

10
Familial Culture, Gender Relationships, and Sexual Indulgence

No one can deny the fact that in the post-1979 era, not only have the material lives of the people of Mainland China greatly improved, especially those in the southern coastal regions, but their horizons have also been widened. In many stories depicted in the first two parts of this book, we can find that the younger the generation, the more individuals know how to arrange their lives autonomously. In fact, even the older generations, men and women in their sixties or seventies alike, have changed their views and lifestyles as the reform and opening-up policy got under way. This seems to suggest that China's efforts in modernization in the past ten to twenty years have indeed made some progress.

We can also say that in Baixiu Village familial culture now has less influence on decisions made in relation to marriage and choice of spouse than it did in the past. Moreover, decision making on daily matters relies more and more on the individual's autonomy and free will. Nevertheless, there are still questions for us to ask, such as: What are the effects of the growth of individual consciousness on gender relationships? Have such relationships been improved? Obviously, there is no simple, straightforward answer to the questions. On one hand, we have observed that respect for one's spouse in two couples who were married in the late 1970s and early 1990s respectively greatly differs from that between Uncle Qiu and his wife, who were married before 1949. We have also seen how Wang Guizhen, who grew up under the influence of the rural economic reform, has planned her own life and direction. This seems to suggest that the social status of women has risen. On the other hand, the industrialization and increase in income have been accompanied by the expansion of some service businesses such as karaoke bars and hair salons, which are used as fronts for the sex trade. The huge influx of out-of-province female workers has also contributed to the prosperity of the sex trade. Many young female workers have yielded to the temptation of easy money and subsequently gone to work in such places. The many young and beautiful girls in the sex trade have therefore attracted many local customers, and the number of local male villagers having extramarital affairs with these girls has markedly risen. These phenomena in turn cause changes in relations between the sexes.

The extreme instrumental values, as well as the pursuit of wealth and sexual gratification, are in people's minds in the entire Pearl River Delta, especially infiltrating everyone in our fieldwork location—Baixiu Village. Some people would say that this phenomenon coexists with the development of any modern city that is steered by a market economy. However, Baixiu Village gives others the impression that the pursuit of wealth and gratification of sexual desires are the main or even the only purpose in life. From our observation, almost all sexually active males have been participating in a variety of sexual activities. It is not uncommon to find principals, senior administrators, and teachers of the village schools who visit such establishments or keep a mistress. There are about two thousand local residents in Baixiu Village, but there are more than twenty thousand female workers from other provinces. People who understand the situation say that besides the sex venues such as karaoke nightclubs or hair salons, it is very likely that 20 to 30 percent of the outside female workers have been selling sex as a part-time occupation or having sexual relationships with local men.

The abundant sex trade in Baixiu Village, coupled with the obsessive pursuit of wealth, present us with a deviant society flooded with corrupted desires and selfishness. At the same time, this may also be one of the motivating forces for Baixiu Village residents. How are these phenomena related to changes in the traditional Chinese concepts of marriage and in gender relationships of the past fifty years? Alternatively, seeing this picture from a wider perspective, do we find that there are certain difficulties present in the course of China's modernization that are reflected by these phenomena?

In chapter 1 of this book, we discussed various viewpoints concerning the women's liberation movement. Many Western scholars who have studied the development of modern China hold the view that women in the Mao

Zedong era had not been truly liberated since the Communist Party gained power in 1949. The father–son relationship remained the axis of the Chinese family and other interpersonal relationships. And the patriarchal culture was still dominating people's lives. But these scholars also point out that when a comparison is drawn between Mao's time and Deng Xiaoping's era after 1979, one could still find more opportunities for women's liberation offered by Mao than by Deng. At least in Mao's time, men and women were mobilized in the same manner to work collectively and to participate in all kinds of political activities. On the contrary, when Deng Xiaoping advocated the market economy and private enterprise, he also hampered the advancement of female social status in many ways. Many women were forced to return home from their workplaces, or were only given work at the lowest levels in factories or enterprises. When consumer businesses arose and the sex trade flourished, women became material or property that could be purchased or owned by others. This materialization of women has become very pervasive. It seems that women have returned to a position of exploitation. Regardless of whether the comparison of Mao's and Deng's times made by Western scholars is correct or not, their analysis of contemporary female social status seems to coincide with the situation in Baixiu Village. Apparently, on one hand, the degree of autonomy enjoyed by Wang Guizhen is much higher than that enjoyed by the women of Mao's time. On the other hand, nowadays most women can only work at very low-level jobs in factories and enterprises. Add to this the fact that women have been extremely commodified by the flourishing sex trade, and all seems to indicate that the present status of women is lower than in the past. Have conflicts surfaced? To explain all these phenomena clearly would not be easy because Baixiu Village is still in the process of evolving.

To put it simply, the sex scene in Baixiu Village can be interpreted as one kind of social abnormality or deviancy. It reflects a decline in the role of familial culture in guiding principles of social behavior. On the other hand, in the dissolution of social order, we find women are being oppressed by patriarchal culture. The oppression-linked phenomenon of women selling sex for money to satisfy men's needs symbolizes that women have become commodities. Thus, while the sex-trade scene signifies the decline in the regulatory power of familial culture, it also strengthens and displays the ideology of patriarchal power dominating women within the context of familial culture.

This part of the book focuses on the theme of sex and pornography, which is also one of the core issues in relationship between the sexes. Thus, if we look at the situation of female Baixiu villagers in the context of Chinese attitude toward sex, we might be able to draw conclusions on how women's status has been affected by the interaction of traditional cultures and the market economy. In this regard, we shall attempt first to outline the attitudes of Chinese people toward sex and examine how this frame of mind affects the sexual behavior of Chinese women. Hopefully, this will provide us with an analytical basis with which to interpret the stories that we are about to uncover in this part of the book.

In Baixiu Village, the relationship between the sexes has been greatly distorted by the attitude of the male villagers toward "sex." This distortion can be seen in the ways the male villagers play with the female body. This is a rare phenomenon in the history and development of the Chinese society. However, if we look at it from a wider perspective, we should not be surprised to find that this phenomenon is the result of traditional values in the Chinese familial culture: "men superior, women inferior." This is also a reflection of the cultural traits of the Chinese people at a deeper level.

Many studies have pointed out, from different angles, that Chinese women are being regarded as the subjects in the realm of sexual behavior, where they are expected to follow the rules governing sex. In other words, only a woman's sexual behavior is subject to assessment in accordance with the standards, but not a man's. [1] Thus, the social norms regarding sex are only used to dictate women's sexual behavior.

Traditional Chinese culture and the Communist Party's social policy have one thing in common and that is the control over the use of terminology in discussion of sexual matters. In turn, this restriction shapes the gender relationship. In a way, it has become a kind of social control.[2] This social phenomenon has been strongly criticized by Western feminists. Although this type of social control has suppressed the development of personal autonomy regardless of the individual's gender, women have been put in a much more difficult position than men. The patriarchal culture, the Communist Party's policy on women, and the private enterprise culture in its early stage of capitalism have all shaped the image of a woman, oppressed women's position in the social hierarchy, and exacerbated the predicament faced by modern Chinese women. Before the traditional values and regulations dissolved, women might still have had peace of mind in that they accepted the position imposed on them by familial culture. But it is already the turn of the twenty-first century; Chinese women have become less and less satisfied with the social position designated for them by familial culture. Yet they retain a relatively humble social standing.

The reform and opening-up have definitely brought some opportunities for Chinese women since 1979. However, ideology and other economical and political factors have not ripened to full maturity. In this case, although women have become more aware of their deprived social status, they have not been able to escape or resolve this hardship. We can see that this has happened to Wang Guizhen and other female villagers. These women clearly understand the effect of the sexual activities in Baixiu Village on their own lives and that these activities are causes of their miseries. But they have no ability to alter the situation. They seem helpless in this respect. Nevertheless, this does not mean that Chinese women now are in a worse situation than before. In fact, in many respects, especially economically speaking, their situation has improved a great deal. It should be emphasized that the female status we are referring to here is considered as having become worse only within the context of the relationship between the sexes. At least this is the situation we find in Baixiu Village. We have observed that female Baixiu villagers are well aware of their own existential predicament and that they exude a kind of sorrowful awareness of the situation.

11

Sex, the Sex Trade, and Extramarital Relationships in the Eyes of Men

When describing the situation in Baixiu Village, compared to other Chinese communities, one might wish to say it is one of "extreme promiscuity." In the eyes of almost every man, any young female could be used to satisfy his sexual desires, and any location that is sheltered or partitioned could be premises on which to perform sexual acts. Attempts to gratify their sexual desires are not only made by adult male villagers such as Zhichao, but also by others of all ages, from elderly men who are over sixty years old to teenagers who may be as young as fifteen or sixteen years old. Every man pursues incessantly and zealously the fulfillment of his sexual desires, just like his passion for wealth. Both pursuits—for monetary wealth and gratification of sexual desires—have become the main goals and motivational forces of the male villagers' lives. Although men are still obligated to take care of their parents, wives, and children materially, they pay much less attention to the emotional needs of their immediate family. Many aspects of relations between the sexes have regressed in an environment where women's bodies have become merely commodities.

In this chapter, we will use real-life stories to illustrate the situation that we have described above. First we will describe the sex scene in Baixiu Village and its neighboring regions. Then we will describe the sexual activities that have taken place in the village and in the primary schools in the vicinity.

We will detail how wealthy principals, senior administrators, and some local teachers seek to have their sexual desires satisfied. Finally, we shall use the stories described in Part I to analyze how men of Baixiu Village look at sex, sex trade, and extramarital relationships.

Sex and the Sex Trade in Baixiu Village

Baixiu Village is under the administration of Qingyang Town. It is situated right along the side of the two-way, three-lane Qingyang Road that runs from Qingyang Town to the county capital. Along the two sides of Qingyang Road and surrounding Baixiu Village, there are four villages. Each of these villages has a population similar to that of Baixiu Village, and each has about 2,000 residing villagers. There are many people from outside, in particular merchants and businessmen from Hong Kong and Taiwan, who have set up factories there. Some of these outsiders employ a great number of out-of-province workers. Take Baixiu Village as an example. In 1994 and 1995, there were between 20,000 and 30,000 workers from out of the province, a large majority of them young women. With the influx of the out-of-province young female workers and the large number of rich and lonely foreign businessmen as well as technical personnel who have been away from home for a long period of time, the village became a breeding ground for the sex trade.

In this region, the sex business takes place in two types of establishments. The first are hair salons, which are of lower class and occupy a smaller area. A hair salon ranges from 20 to 70 square meters in size and employs several to a dozen or so female workers, all of whom are from other provinces. Besides offering hairdressing services, they provide "sex services" as well. Usually there are smaller rooms annexed to the main salon, which are used for these sexual transactions. This kind of service costs between 50 and 100 yuan. According to some unofficial statistics, there are approximately fifty or so salons in Baixiu Village and its vicinity.

The second type of establishment is the karaoke nightclub. The premises are much larger than those of a hair salon. In Baixiu Village and its vicinity, a karaoke nightclub is from 100 to 500 square meters in size. There are some karaokes as large as over 2,000 square meters in Qingyang Town. The nightclubs in Baixiu Village and its vicinity have deluxe equipment and furnishings. The "sing-along girls" hired to work in the karaoke nightclubs are more beautiful and better dressed than the hair salon workers, and karaoke fees are higher than those of hair salons. Most of these nightclubs are located along Qingyang Road, where there are twenty or so nightclubs and bars.

The target customers of the hair salons are the lower-income local villagers, whereas the nightclub customers are the employers, senior managers, and technicians from Hong Kong and Taiwan, as well as the local residents of higher income.

Similar to other villages and towns in the Pearl River Delta, the sex trade in Baixiu Village and its vicinity did not commence until the late 1980s. But in only a few years it flourished, reaching its peak in the early 1990s. Then it began going downhill.

There was not a single sex den in Baixiu Village and its neighborhood before the late 1980s. Nor could people there afford such entertainment. After 1979 and in the early stage of social reform, the situation in this respect remained unchanged—that is, there were no sex establishments. Later, there were a few restaurant-type places for social gatherings. At night, people hung out at these places to eat and drink. At times, there might be some attractive women from other provinces serving these customers. These female waitresses soon began talking and flirting with the customers. With the development of the market economy, the sex business also took off.

The first karaoke nightclub was opened in Qingyang Town in the late 1980s. Its major customers are the businessmen and management staff who came from Hong Kong and Taiwan to establish factories in this region. Within several years, the whole town was filled with these sex joints disguised as entertainment establishments. At the same time, these places become the major leisure and entertainment centers for local men.

This business was at its prime in the early 1990s. At that time, dressed-up women and shady-looking men could be seen everywhere in the evenings from seven o'clock until past midnight. The consumption and spending in this area had spread to other businesses, such as restaurants, retail fashion, and taxi services, which also became much busier. But this did not translate into good business all the time. Whenever there was government action to squash vice, these other "related" businesses would also become quiet. There were several anti-sex trade government actions each year. The most severe action took place between late 1993 and early 1994. During this time, any men or women who went to karaoke nightclubs even just to sing songs were also questioned. Many of these nightclubs closed permanently or temporarily because of the government's actions. The whole of Qingyang Town went into a kind of economic depression, and many male villagers felt a sense of loss.

This kind of feeling can be easily understood because Qingyang Town had been transformed from a quiet, simple rural village into a vibrant colorful world of song and dance within just a few years. The supply of young, beautiful sing-along girls appeared endless. It seemed as if these girls had been sent to satisfy the basic instincts of men that had been suppressed for a very long time. In fact, many men who came from Hong Kong and Taiwan shared the same feeling. However, the extreme indulgence in sexual pleasure began to subside in Baixiu Village and its vicinity in the middle of 1994. Karaoke nightclubs are still the major entertainment centers for the male villagers, but the scene is not as feverish as before. This, on the one hand, is because the economic development of Baixiu Village has slowed down. On the other hand, it is because the business has lost a great deal of its novelty. However, this does not mean that men are less involved in extramarital relationships. In fact, since the late 1980s, besides visiting the karaoke nightclubs, Baixiu Village men would also turn to the outside female workers to have their sexual desires satisfied.

Insiders told us that among the 20,000 or so outside female workers, 20 to 30 percent have had sexual relationships with local men. These relationships might be one-time transactions, or transform from short-term to long-term. In fact, many nightclub or salon girls were originally factory workers; some might have been recruited directly from their hometowns. When the business was at its peak, some nightclub owners sent representatives to other provinces to recruit sing-along girls. But female workers have always been the main source of recruitment for this business.

In Baixiu Village, the reason that factory female workers become "service girls" is twofold: economic need and peer influence. From the early 1990s until now, the wages of factory workers have not been increased. But the standard of living has inflated a great deal. A factory worker at entry level would get 300 yuan a month and an overtime bonus of 3 yuan per evening. This adds up to approximately a monthly wage of 400 yuan. From this, the employer would deduct 150 yuan for room and board. A worker would spend about 100 yuan on other miscellaneous living expenses. That leaves about 150 yuan of disposable income. But in the Pearl River Delta, where Baixiu Village is situated, the cost of living is very high compared to that of other regions of China. For example, it costs 2 to 3 yuan for a motorcycle ride from one end of Baixiu Village to the industrial area, and the trip only takes about three minutes. Meals provided by the factories are of poor quality and do not provide workers with sufficient nutrition. Many times, workers must eat elsewhere. Meals at a small restaurant would cost 3 to 5 yuan. Most workers did not have any recreational activity during holidays because they could not afford it.

Of course, when compared to the living standards in most of the female workers' hometowns, the life in Baixiu Village was far superior. Otherwise, they would not have been enticed to leave home and go to work there. However, this contrast in the living standards makes materialism even more attractive. Each night these service girls, just by spending a few hours singing, drinking, and sometimes flirting with the customers, would get a gratuity ranging from 100 to 300 yuan. They would get more if they were willing to have sex with the customers. It was almost impossible for the factory workers who earned such meager wages not to think of becoming one of them. That is also why some better-looking workers were willing to have sexual relationships with their supervisors in exchange for some material benefits, which could be in the form of a promotion and a raise in salary, or a generous bonus or allowance. Others having similar or better physical qualities would then follow suit. In effect, woman workers who were quite attractive were almost destined to become one of these girls. This explains why 20 to 30 percent of the female workers have had sexual transactions of one kind or another with Hong Kong, Taiwanese, or local men.

One could say that it is very common to find men and women in Baixiu Village having extramarital sexual relationships. This does not mean "sexual liberation"; it is better described as "promiscuity." However, this description does not apply to the local women. These women continue to hold and preserve the traditional viewpoints and conservative attitude toward sex. Even the promiscuous female workers and nightclub girls, who are treated by others and even themselves with contempt for their sexual behavior, hold these same views. Thus, one could say that Baixiu Village is a male-dominated community where men are free of any restraints, moral or otherwise, to have their sexual desires satisfied. The following stories illustrate this phenomenon.

The Extramarital Sexual Behavior of Schoolteachers

It was approaching dusk on an autumn afternoon in 1994 when I went to Baixiu Village Primary School hoping to interview some teachers there. Baixiu Village Primary School occupies a large area. It was established in 1987. The building cost was about 800,000 yuan. At that time, this was a very large sum for a public expenditure. It would not have been possible without the rural economic reform and opening up. In fact, the buildings of the primary school have a rather modern appearance, but they lack repairs and maintenance, and look rather rundown now. In China, this is a very common phenomenon in buildings several years after they are built.

I went into the school's main entrance. It was very quiet and no one was in sight, probably because classes were finished for the day. In the corridor, I ran into a young woman in her mid-twenties. I thought she was one of the teachers and told her the purpose of my visit. She thought for a moment and said that all teachers had gone home except one—Mr. Wang, who came from

Guangxi Province. He boarded at school and therefore might be available for an interview. She led me to Mr. Wang's room, which was located next to the stairs. Mr. Wang answered the door. He appeared to be about thirty years old, 160 centimeters tall, and of medium build. He was wearing a T-shirt and shorts. He looked friendly and invited me into his room.

The boarding room was approximately 7 square meters in size. There was a bunk bed on one side and a wooden desk and two chairs on the other. Mr. Wang was preparing his dinner. The food was simple. He used an electric rice cooker to make rice and steam vegetables and meat.

Mr. Wang told me that he was originally from Guangxi and was thirty-one years old. He had been a teacher in his hometown for over ten years. He was introduced through some friends to the teaching position at Baixiu Village Primary School. He left his wife and two children to take up teaching at Baixiu Primary since the beginning of this school term. Mr. Wang was content with his present salary, which earned him a total of slightly more than 900 yuan a month including a bonus and cash rewards. Although his income was from 200 to 500 yuan less than what the local teachers earn, it was much higher than what he made in his hometown, where he could only get 250 yuan a month.

My first conversation with Mr. Wang covered only very general topics. Since it was our first meeting, Mr. Wang, though talkative, spoke rather cautiously. Later, each time I went to Baixiu Village, I would visit him and invite him to dinner. We gradually knew each other better and our discussion became more in-depth.

During some of our conversations, Mr. Wang expressed some dissatisfaction with the interpersonal relationships at the school as well as students' learning attitude. In particular, he was discontented with students' manner toward their teachers. He said everything was "money first" in this place. Everyone—men or women, young or old—used wealth to measure his or her status and identity. Accompanying these extreme instrumental values were the efforts spent on having one's sexual urges satisfied. In Mr. Wang's opinion, there was no man in Baixiu Village who was not fond of philandering. Mr. Wang knew that some older boys, who were only in their primary five or six, about ten to thirteen years old, especially those who had failed and remained at the same grade level, would go to the karaoke nightclubs and seek the company of sing-along girls.

As for the schoolteachers, most local ones were regular customers of the nightclubs. Mr. Wang pointed out that the principal was one of the most frequent customers. He told me that the young woman who led me to his room was one of the many mistresses of the principal, in addition to her regular duty at school, which was cooking for the teachers. As I hoped to obtain more details of the situation, I begged Mr. Wang to arrange for an interview with the principal.

One morning, Mr. Wang led me to the principal's office for an appointment with the head of Baixiu Village Primary School. The principal and the senior master of curriculum shared the same office. It so happened that the senior master was also in the office. Both were local people. They were of the same age, that is, about fifty years old. The senior master had a smaller build and looked very timid. The principal had a medium to slightly heavy physique. At first glance, his appearance and outfit gave the impression that he was a gangster, or a person who conducted illegal activities.

After I entered the office, the principal gestured for me to sit down on a chair next to him. Then he ordered Mr. Wang to pour me a cup of tea. His order was almost in the form of a command. His tone startled me. After having sat down, the principal looked at me from the corner of his eyes and said, "So you are the person who is doing research in the village?" His tone was condescending and disrespectful.

I nodded and said "yes." He then continued, "What is the use of it? And how much can you make?"

I did not know how to respond because I was surprised at the questions posed by a person of his status under such circumstances, and also because I remembered that Mr. Wang had told me that the principal had an excellent relationship with the party branch secretary of Baixiu Village. I thought if my response offended the principal, it would affect my subsequent research work at the village. I briefly and respectfully explained our study. Then I asked him about the development of Baixiu Village in recent years. He said that everything had been progressing well. Then he told me his own story of success.

"I am the principal of this school and earn about 1,500 yuan a month. But this is only a small part of my income. I own factories in Guangzhou and its neighboring districts. I have over 2 million yuan worth of assets, far more than what my relatives or friends in Hong Kong have."

He then pointed to a minivan parked not too far outside the window and said, "That is the vehicle I drive every day to school." He paused for a moment and continued, "Actually I don't come to school on a full-time basis. Whenever it's necessary, I can leave."

The principal boasted about himself continuously during our meeting. When he talked about the sex scene and gender relationships in the village, at first he stressed that the relationship between men and women was normal and that most families were living harmoniously. However, when I repeatedly asked whether the booming sex business would affect men and other interpersonal relationships, he then revealed some truths.

"Yes, there have been some individual incidents. But there is no man who would not be playful at times! But to be a wife, she must understand the situation. They usually do not have any complaints."

Some Baixiu villagers later told me that everyone in the village knew that the principal had been keeping more than one mistress. They also recounted an incident that illustrated how far the principal would go to get sexual enjoyment.

Every year, the schoolteachers take the graduating class to a neighboring county for a sleepover journey. The group stays in hotels, with one teacher sharing a room with several students. Travelers who stay in the hotels in the Pearl River Delta often receive phone calls asking if the male guests require some "service girls" to accompany them "to go to bed." At about seven o'clock in the evening on the night of one such sleepover journey, the Baixiu Village principal, who was staying with several students in a hotel room, received such a phone call. He told the students that he needed to attend to some business and asked them to leave the room for a couple of hours. The students left the room as told. But they did not stay away for very long, and wanted to return to their hotel room. They knocked on the door for a long time but the principal did not answer. So they went to ask the concierge to open the door. But the door was chained up from inside. A moment later, the principal came out with a young woman who had seemingly dressed in a great hurry.

It is very common for principals, senior administrators, and teachers to keep mistresses and frequent prostitutes. But it is rare for people like the principal of Baixiu Village Primary School to act so obviously and openly. I have also interviewed the principal of a primary school of a neighboring village. He was in his fifties and looked quite decent. After the interview and on my way home, I said to my companion who had gone with me for the meeting that this principal did not look like he was fooling around with women. But my companion, who knew the village very well, said that I was naive. This principal was another philanderer and sexually promiscuous. He often had "service girls" go to his office to have sex with him during lunch breaks. My companion also told me that a senior administrator of that same primary school had been keeping mistresses. There was one time when the wife of the senior administrator and members of the local security committee attacked the mistress and ransacked her home.

According to Chinese tradition, moral standards as well as social norms, the "teacher" plays not only the role of master transmitting knowledge to his apprentices, but is also the role model for students' moral character and integrity. Among all virtues, it is most important that a teacher should be very strict about his or her relationship with the opposite gender. Obviously, this traditional demand of such ethical standards from a "teacher" in modern

China is no longer being upheld. But to most Chinese people, especially those who live in the rural areas of Mainland China, as opposed to other Chinese who live in overseas communities, the respect for "teacher" is still there. This can be seen from the respect for our research-team members accorded by most villagers. In fact, when we conducted interviews with principals and teachers of the primary schools, they criticized the phenomenon of sex trade and promiscuity in the villages, although they might not have genuinely meant it. Many of them indicated that they were abiding by the traditional values and ethical standards, that they always emphasized the importance of harmony within the family, and that they led a clean and chaste life. Take the example of Mr. Wang. When talking about the sexual activities of other school teachers, he often showed his aversion and distaste toward the behavior of his colleagues. But when I invited him to have dinner at some high-class restaurants, he would act in front of the young female waitresses as if he were a big spender who often went to these high-class places.

Mr. Wang told me of another incident. An older, wealthy student who had failed and had to remain in the same grade had once boastfully said to Mr. Wang that a month's salary of a teacher was not enough for him to entertain himself for a night or two at a karaoke nightclub. The student had invited Mr. Wang to go to a nightclub so that he could see for himself the new experience. Other teachers also teased him for never having gone to the hair salons or nightclubs. He said angrily, "Not long ago I contracted for an orchard at my hometown. Next year there should be a good harvest. Then I will be able to invite these fellows to the nightclubs and have some sing-along girls serve them. I wonder if these people will look down on me ever again!"

This reaction illustrates a classic example of the mismatch between "belief" and "behavior." It is true that the male school teachers are also men. They are just like any other man of the village who may not be able to resist the temptation of sexual desire. But to have found someone who fools around with women as shamelessly as the principal of Baixiu Village Primary School does reflect the seriousness of men's obsession with sex in this area. At the same time, this phenomenon illustrates the fact that women have been turned into commodities. The patriarchal culture of familism has reemerged in another form and seriously distorted the relationship between the sexes. In the next two sections, I will use the perspective of other male villagers to look at the situation.

The Feelings of the Older Male Villagers

"Sex" has long been taboo to Chinese people. It should never be a topic for open discussion; this is even more so among family members. Any businesses or activities that are related to "sex" appear in an "asexual" format.

For example, the premises of a business that obviously provides sexual services would be called by names that have nothing to do with sex, such as "hair salon," or "karaoke nightclub."[1] In the previous section, we described how the principal and schoolteachers of Baixiu Primary School were obsessed with sex. But in public, especially in front of strangers, they would deny participation in any of these sexual activities. Some might even criticize the phenomenon openly. This approach of "asexualizing" the "sexual businesses" illustrates the "father–son dyad" form of gender relationship; it also strengthens the patriarchal culture of familism and further distorts the already unequal gender relationship in which women become more oppressed. Now, we shall unveil the attitude of the older-generation male villagers toward sex.

When I was with Uncle Qiu, we seldom discussed the sexual activities in the village. But I had always wanted to know what men of Uncle Qiu's generation thought about them. One evening, he came to my place as usual to watch television and chitchat. I asked him about the sex situation in the village. Surprisingly, he said he did not know anything about it. It seemed that he was being evasive, trying to avoid answering the question.

"Ay, I don't really know these things." He sighed and went on, "Besides, I don't really care if other people have been 'playing with women.' That's their own business." Then he stopped and said no more. He just sat there and smoked. It seemed that he wanted to change the topic to things such as criticizing the television programs. This was completely different from other times when we discussed other matters of the village. At those times, he had been very talkative. I became more curious and kept on asking. Finally, I discerned his opinion, which is as follows.

"There are three most corrupt things in any society. They are sex, gambling, and drugs. These are the bad fruits of reform and opening-up. These things would not have happened in Mao Zedong's time. In my opinion, the central government should start cleaning up from the very top level. These so-called anti-vice campaigns now are useless. The lower levels do not care. But if they begin these efforts from the very top, the whole phenomenon may be uprooted and the entire scene could be changed."

Again, he did not want to answer my questions directly, but criticized the government policies. Although this is his usual way of expressing his opinions, often side-stepping a great deal, this time it looked like he did not want to discuss this particular question. Since I repeatedly asked him about the situation, he slowly revealed some of his thinking: "If men who do not have a wife find a woman to fulfill their needs, I don't see a problem. This is basic human nature and normal reaction. But if these men who already have a wife still 'play with other women,' this obviously is not good because that would break up the harmony of a family. Sometimes, some women even want to pick up an ax and hack up their husbands. Ah! To get involved in these things is really bad. But nowadays all young men become obsessed with these activities. These are social phenomena that are hard to judge and I really don't understand them. I think these are things caught from you Hong Kong people. There was no similar set-up here in the past." He paused for a moment and continued, "In the past, even those men who had several wives would not go outside and 'play with other women.'"

"Have you ever 'played with women?' Have you ever thought of doing so?" I asked jokingly.

He paused for a long while before he said, "Ay! You really think it is that easy? Where would I get the kind of money to 'play with women,' 'keep mistresses'?" When he finished the sentence, he fell into deep thought. After a while he left my place without further comment. In the past, after I finished interviewing him he would have stayed for hours and chatted with me. But this time was an exception. Perhaps this happened to be a topic that made him feel sad. He looked melancholy when he left.

The second day, I met Uncle Min at Uncle Qiu's barber shop. Uncle Min often likes to contradict and challenge Uncle Qiu. So I thought that if I reopened the previous evening's topic, it might stir up a heated argument, which might reveal their true thoughts. So I brought up the recent anti-vice campaign to see what their reactions were. Uncle Qiu, as before, said he was not familiar with the topic. At first Uncle Min did not express any opinion directly and sidetracked the discussion. Then he talked about some things related to the campaign. But his view on the topic was still obscure. He did not say whether it was right or wrong of men to "play with women." His words were equivocal.

Uncle Min then mentioned where in the vicinity of Baixiu Village one could find the "ladies." I pretended that I was interested and asked, "So you can find them in this village too?" Uncle Qiu responded very quickly, "There is one hair salon right across the ancestral burial ground. You can also find another hair salon in the village." Obviously, he knows the locations of all the sexual services in the village.

It seemed that the discussion had begun to boil. Uncle Min then expressed his point of view: "Actually there are many people, from young to old, who 'play with women' in the village. Rich men would even 'contract' these women on a long-term basis."

"Ay! This makes for a social problem!" Uncle Qiu said regretfully, shaking his head.

"I don't see any problem for the rich guys to keep 'ladies.' If you go out to play often, today this lady, tomorrow that, that is even worse," Uncle Min contradicted Uncle Qiu's viewpoint.

"If you act like that you create chaos in the family. This is the problem. It is decadent. It is illegal. If you don't have a wife, you don't have another solution for your sexual needs. Then it is normal for men to have those needs. But for those who have a wife, this would cause conflict in the family. That is bad," Uncle Qiu said angrily.

Uncle Min immediately responded, "It was just like the past: A man could have several wives. Our ancestor Jianqiang was the offspring of a second wife. It is only in the present time that we have the institution of monogamy, which does not allow one to have two or more wives."

"That is why it is illegal to have two wives. To be illegal is to be decadent," Uncle Qiu said loudly.

Uncle Min argued, "Look, you are talking nonsense again. How can you say to be illegal is to be decadent? It may be so with bribery and drug addiction, but definitely not in the case of having one more wife."

"If you don't have money, but use the 'household expenses' to 'play with women' or keep mistresses, wouldn't this cause disharmony to the family? If you don't have any money, you would use every means and effort to get it, for example to bribe or accept bribes. Wouldn't it become decadent?" Uncle Qiu said rather arbitrarily.

Uncle Min said, "You go 'play with women' and do not give money to your wife, she would certainly become angry. She would quarrel with you or the two of you might even resort to physical violence. Or if you are rich, but you only give money to your second wife, I assure you that there would be fighting between the couple. But if you are rich and give 'household expenses' to your first wife, and still you have extra money to keep mistresses, I don't see any problem. If you have money to give to your women to spend, why would they want to bother you? Only unreasonable women, like those tyrannical Hong Kong women who have a lot of money to spend, would still try to stop the husbands from 'playing with women' and keeping mistresses. The women here generally would not act like that. If you have money for her, she would not care if you keep a mistress or not."

"For sure you are right about the rich guys. But there would definitely be trouble for those who don't have the money." Uncle Qiu seemed weakened in his response and began to agree with Uncle Min's viewpoint.

Our conversation ended at this point and an interesting picture emerged. Throughout the conversation, Uncle Qiu and Uncle Min looked as if they had each taken a different stance, opposite to that of the other. But when we examine it more closely, their viewpoints begin to converge. Because Uncle Min often likes to contradict what Uncle Qiu says, they give the false impression to other people that they hold opposite positions. In fact, both men think that the ultimate issue in the extramarital sex question is one of money.

Perhaps the only difference is that one is poorer and the other richer. Uncle Qiu lives more stringently whereas Uncle Min is much better off. They agree that if a man were wealthy, it would not be an issue that he "plays with women" or keeps mistresses.

We also see that at first Uncle Min was unwilling to talk about this subject. I have seen h m on other occasions, like Uncle Qiu, reprimanding the young people for indulging in sexual pleasures, calling it decadent behavior. But in reality, Uncle Min is one among the older generation who rather enjoys this type of activity. Zhichao told me that Uncle Min procures sexual services of female factory workers for men of his generation. Zhichao and others have often used their motorcycles to transport the female workers to a destination where Uncle Min would be waiting and receiving the girls. Afterward, the drivers would go to the same place to pick up these girls and take them back.

As for Uncle Qiu, one could be fooled by his apparent distaste for other men who "play with women." Although he did openly reprimand such decadence, he enjoyed staring at women himself. In fact, he is also well-acquainted with some "ladies." Someone told me that several days a month he would dress up and go to Qingyang Town to visit these "ladies."

Here we can see the attitudes of the older-generation men toward extramarital sexual relationships. We are also able to perceive women's social status through men's attitudes. The open reprimand of sexual indulgence has, to a certain extent, moral implication. But we have to be careful here; these old men's reproachful attitudes toward sexual indulgence has nothing to do with fidelity and obligation toward wives. Rather it implies that sex itself is a taboo that should not be touched upon openly. As mentioned earlier, it should be observed strictly that any activities that are related to "sex" should be cloaked as "asexual." Thus, it has evolved that sex is regarded as moral taboo.

From the perspectives of traditional family culture and Francis Hsu's theory of "father–son dyad," sexuality as a moral taboo further suggests the nature of gender inequality regarding "asexuality" in the context of male-dominant relationship. To men, it is a traditional belief that "sex" is a moral taboo for them to talk about, yet this traditional taboo does not extend to their deeds. A man can have as many wives and concubines as he wants. During the Song and Ming dynasties, moralists could still frequent brothels and would not attract moral criticism.[2] That is why Uncle Min and Uncle Qiu would think that as long as husbands can satisfy their wives' needs for money they can "play with women" and keep mistresses without attracting any criticism.

In fact, all this reveals the degree to which women are being oppressed. To women, "sex," being taboo, has a meaning in moral terms that is exactly the opposite of that for men. Sexual prohibition is closely tied to actual sexual behavior and women's fidelity to men. So-called "asexual" sex is the standard used to measure the propriety of women's sexual behavior and their sexual relationships. If the main purpose of women's sexual activity did not focus on reproduction, they would be in violation of the standards of "asexual" sex. Their behavior would then attract moral criticism. Obviously, above all, women's extramarital sexual activity is considered most reproachful because this type of relationship shows that the main purpose of sex is for enjoyment.

The issues in relation to women will be discussed in detail in the next chapter. Here we have just shown that the basic primal sexual desires of older-generation men have been released by the improvement in material life brought about by the reform and opening-up policy together with other external factors. However, men of the older generation have been deeply influenced by familial culture, yet almost untouched by social values affiliated with the market economy. Moreover, the development of an individual consciousness of the self in these older men is not as advanced as that of the younger generation. They also lack respect for other people's autonomy. All this, in addition to their obsessive pursuit of sexual pleasure, strengthens their ideas of patriarchal culture and reenforces the unequal gender relationship that has been rooted in their minds.

Sexual Desires of the Younger Male Villagers

Although Zhichao's choice of a wife was his own, and not arranged by parents, he does not seem to feel satisfied with the status quo. He told me many times that he hoped he would have the opportunity in the future to marry one more wife.

"Several years ago, a fortune-teller said that I would have two wives. He wouldn't lie to me. Although the government regulations only allow one wife, there are many people in the village now who secretly have a second woman. Doesn't that mean two wives?"

"Why do you want to have one more wife? Don't you like Chao Sao anymore? You married her of your own free will," I said.

"I have told you before. It was after I had decided to get married that I realized I did not want to marry her. But because we had dated for two years, I felt sorry for her and did not want to abandon her just like that. It would have been very hard on her. Maybe because I am a very caring person!

"I think what that fortune-teller said is probably true. I find that a lot of women like me. Even those 'service ladies' sometimes approach me and say that they would not take money from me for having sex with them. There are just too many who like me. But most of these relationships are fruitless." He began to tell me of his extramarital relationships and sexual experiences.

"When I was selling salted preserved fish in Qingyang Town, there was a woman who liked me a lot. She often came to the fish store to see me. She did not believe that I was married. She went to the temple and asked for an oracle. She wanted to see if there was any chance of marrying me. One time she got an excellent marriage oracle. She immediately came to tell me about it. At that time, almost all my neighbors around the fish store knew that she liked me a lot. In fact, she was quite good-looking. But I had to keep a distance from her. I did not want our relationship to go any further because she was a local resident. If she were just looking for pleasure, enjoyment, I would see no problem having sex with her. But since she was serious about it, it would be bad if things got out of control.

"Because she did not believe that I was already married, she often asked me to take her home to meet my parents. So I decided to invite her to come to my house and let her see that I have a wife at home. She really came. When she saw my wife, she was dumbfounded. I told her, 'Say hi to Ah Sao [sister-in-law]!' She finally accepted the fact that I am a married man and left me in peace. If she were not a local, I would definitely have kept dating her to see what would happen next."

Before he worked for the fish store, Zhichao had been a laborer in a factory in Baixiu Village. At that time, he had already had sexual relationships with female factory workers. According to what he said, since they were all from out of province, it would be no big deal to have sexual relationships with these women. He even thought that these women really enjoyed it.

"When I was working as a laborer at the factory, an 'out-of-province girl' was very friendly with me. One night when I was moving cargo in the warehouse, she came in suddenly and held me close. She wasn't pretty at all. Besides, she was very nearsighted and wore glasses. But she made me very excited. We then had sexual intercourse inside the factory warehouse. The next day the 'out-of-province girl,' who must have regretted the whole thing, wrote to me and asked me to forget the incident that night. She said that she had been too impulsive. But not long afterward we had sex several more times. She worked at the factory for over a year. When she left, I gave her 200 yuan. Similar situations have happened to me several times."

"Are you not afraid that Chao Sao will find out about them?"

"My wife really trusts me. She thinks that I would never go 'playing with women.' Remember I took home the girl from town? Even then she was not suspicious of me. The only thing that she is afraid of is that I go gambling. But she has never asked me if I 'play with women.'

"In fact, it is inevitable that men sometimes go 'play with women.' But when they do, they should be careful not to get involved, not to let their wives know of it. Then it should not be a problem. That is why I would not want to get involved with the woman I met at the fish store. Because once the word got out, I would be in deep trouble. Sometimes people get bored with living with the same woman for too long," he said matter-of-factly. "It is common for men to seek pleasure. But one should never get carried away. Always be nice to the wife. Then there should not be a problem."

Zhichao always boasted that he never spent money on "playing with women." In fact, many women took the initiative and offered themselves to him.

"The 'out-of-province girl' who works in the hair salon across from the restaurant flirts with me every time I pass by her workplace. She invites me to go to the small room upstairs. She always says that she want to have sex with me even though I don't give her money. There are many factory girls who want me. Sometimes I introduce the 'northern guys' [men who came from the northern provinces to work at the factory]. Then they always want to buy me a gift. Of course I don't want to accept that because their wages are really low. So they often introduce some factory girls to me in return. Of course it is free of charge." Zhichao was holding a cigarette, smiling with eyes half shut, apparently enjoying talking about his sexual experience.

"I really fooled around with many factory girls before. Ten, twenty girls a month. Women are just like soda pop cans and chewing gum to me. When I'm finished drinking and chewing, I just throw the can away or spit the gum out. But ever since I recovered after hurting my leg last year, my health has not been as good. That's why I play around less now."

When Zhichao had no customers, he often sat under the banyan tree in the marketplace and chatted with other motorcycle drivers. Most of the time, the topic was nothing but women. They would talk, for example, about which karaoke nightclub has the most sing-along girls, which one has the prettiest or newest batch of girls. Sometimes they would describe the women they had slept with, telling the others about their figures and their performance in bed. Sometimes, when they got excited by their own talk, they would start flirting with factory girls who happened to pass by. I witnessed one such incident. There was an outside factory worker with whom they were acquainted. She wanted a ride back to the factory. She was quite pretty. I saw Zhichao use his hand to stroke her face and say flirtatiously, "This feels really soft and smooth!" The worker was obviously upset and pushed Zhichao's hand aside. All the other drivers were amused and laughed heartily.

In fact, Zhichao wishes more for a second wife than just to have these short-term sexual relationships. Besides the effect of the fortune-teller's words, he saw many rich men in the village have "bought" second wives. He saw, too, that his two younger brothers who were factory managers also had second wives. He believes that some day he will have the opportunity to do so. But so far he has not met a woman he likes. More important, he does not have that kind of money yet. He does not think that to "buy a second wife" would be unfaithful to Chao Sao. It is a very common thing. To him, it would be strange if men did not do so or want to do so.

"Nowadays, it is very common for men to have two wives. Some even have more than two. Wasn't it true that people in the past did the same? The rich people of our grandfather's generation usually had two wives or more. I don't see any problem in having two wives. It is more important for men to have the money to provide sufficient living expenses for both their first and second wives. One cannot just give money to his second wife and ignore the first. This is wrong. But if you are rich, it should not be a problem if you contract more wives. I don't see why the first wives should complain about husbands doing these things. I have no pity for those wives. I only buy a second wife. I am not neglecting my first wife. I am not cutting her off from her living expenses."

The generation of middle-aged men like Zhichao basically share the same viewpoint as the older generation, such as Uncle Qiu and Uncle Min, perhaps even more capriciously and with greater abandon. In fact, even men of a younger generation, like Qiming and others, have similar thoughts.

One afternoon, I was having a beer in a small restaurant in the village with Qiming and Ah Song, who are both the same age. While we were drinking, we talked about the question of extramarital relationships. Qiming told me that he also had "played with women." He believed that almost every man in Baixiu Village did so.

"I also go 'play with women.' Sometimes, a bunch of friends go to a karaoke nightclub. Everyone is excited and happy. Everyone joins in the spirit of the occasion. You know, what man would not want to 'play with women?' We have a lifetime ahead of us; how can we stay with the same wife our whole life? It would become stale. But just don't go overboard. Like me, I only 'play with women' once in a while. It should not be a problem. As for 'buying second wives,' it would be bad if you neglect your wife! It is infidelity!"

Ah Song agreed with Qiming. He was not yet married. Almost every night he went to nightclubs. He thought it was impossible for men not to "play with women" at all. As long as it did not become an obsession, he saw no problem in doing so.

"What do your parents think of you 'playing with women?'"

"A reasonable parent will not ask. In fact many men in the past had three or four wives," Ah Song replied.

"Although you do not do it too often, what would happen if your wife found out?" I asked Qiming.

"How would she know if you do not tell her? Since the wife doesn't know, why would there be any problem? If she finds out, then there would definitely be serious conflicts between the couple. But I think they would be no more than just quarrels. Generally speaking, the village women try to put up with their husbands' activities in this regard. Especially if husbands provide the wives with sufficient money, the wives would not become rebellious. Some wives would turn a blind eye to these things. Some even move to the town and pretend they know nothing about it," Qiming said.

"In reality, are there any couples who have divorced because the husbands 'played with women' or 'bought second wives'?"

Qiming replied, "I don't think there has been any such incident! Women would not divorce their husbands. They would try to tolerate things. To get a divorce here is not that simple. Divorce will attract a lot of gossip. They would not be able to stand it. So the worst case would be to make a scene and quarrel with their husbands."

It is clear that men of different generations in Baixiu Village share the same views toward extramarital relationships and sexual activities. Uncle Qiu and Uncle Min share the same outlook. Zhichao and Qiming hold no different viewpoints. The only difference is that the former do not care about their wives' feelings, yet the latter are a little worried if their wives find out. Most marriages of the older generation were arranged by the parents and matchmakers; these marriages are built on a functional basis, rather than on love between the couple. Besides, the concept of "men superior, women inferior" embedded in the familial culture has been deeply rooted in the minds of people of this older generation, thus making this generation accept the phenomenon of "buying second wives" more readily. In other words, men of the older generation feel more at ease and guilt-free with the activity of buying second wives than their younger counterparts. However, it is also the men of the older generation who are very reluctant to discuss openly the sexual activities in the village and their own extramarital relationships. How do we account for this contradiction?

The data we collected from the fieldwork indicate that not many village men aged sixty or more "buy a second wife." One of the reasons is that they are not the biggest beneficiaries of the reform and opening-up policy. Probably due to their age, they are less adapted to the pace of rural economic reform as well as less capable of finding profitable means. Thus they have less money to maintain the lifestyle of possessing a second wife. Generally, they look for short-term relationships in order to have their sexual desires gratified. The act of frequenting prostitutes is looked down on by Chinese traditional morals. Moreover, "buying second wives" is different from marrying concubines in the old days. The latter constituted a legal marriage and was done in public. However, "buying second wives" is an act done secretly. It is also illegal under the present laws. Legal sanction combined with the traditional concept of sex being a taboo may, to a certain extent, explain this contradiction. As mentioned in the last section, the unwillingness to discuss this taboo in the open as well as the guilty feelings and censure associated with the moral standards are not founded on the sense of infidelity to wife and family. Instead, men are mainly restrained by the traditional concept that such acts are a kind of taboo, especially since they are also illegal.

In an interview with the owner of a mini karaoke nightclub, I asked why no local people would "play with women" during the anti-vice campaigns. My initial thought was why should men be scared? If any were caught, they would at most be penalized a few thousand yuan and sent to jail for a couple of weeks. The nightclub owner was so appalled by my question that his eyes opened wide and he said, "No, this is unthinkable! Their reputation would be ruined completely." But then everyone in the village already knows that almost every man has extramarital sexual relationships. These adulterers are not ashamed of such behavior, so how could their reputations be ruined? Does it refer to the act of frequenting brothels being made known? Or does it mean they are involved in a criminal act? This remains unclear.

There is another intriguing example of the moral murkiness surrounding adultery. The home of the second wife of the senior administrator who taught in the primary school next to Baixiu Village was attacked and ransacked by the administrator's wife, mother-in-law, and members of the security committee. They also beat up the second wife. Since it was an internal affair of the village, no one was fined or jailed or punished by law. But this incident was well known in the area. Ever since it happened, the primary school administrator felt very ashamed and could not hold his head up. It is obvious that this shame was not derived from any legal sanction, nor was it caused by his sense of guilt for having betrayed his wife. Perhaps it can only be explained by the fact that sex has long been suppressed and regarded as taboo, that people are banned from discussing it in the open, and that sexual behavior is still not acknowledged in public.

In effect, these cases reflect how familial culture has shaped the characteristics and self-concept of the Chinese people. In the discussion of the theoretical framework in the first chapter, we pointed out that Chinese people do not behave according to individual feelings. The "individual self" is appended to the "larger self." Individuals are submissive to society. In matters involving value orientation, especially in the public arena, the social norms projected from the "larger self" must be used as the behavioral standard for individuals. Social norms further suppress individual sentiments. At least, people do not express their inner feelings in public. In this respect, if an individual is reprimanded by the public, he or she would feel a loss of "face."[3]

On the other hand, to men of the younger generations like Zhichao and Qiming, the idea of sex as a taboo not to be discussed openly has much less weight. They are not afraid of acknowledging their extramarital sexual behavior. They stress the point that as long as they provide their wives with sufficient money, they have the right to buy second wives. Nevertheless, since Qiming's generation is more advanced in the development of individual autonomy, they have more respect for other people's choices. In addition, most of their marriages are based on love and passion. They often try not to let their wives learn of their extramarital sexual activities. And they feel more of a sense of guilt about betraying their wives. In conclusion, the "father–son dyad" type of interpersonal relationship is still the main basis of the relationship between the sexes in Baixiu Village.

12

Wealth, Materialization, and the Feminine Viewpoint

It is easy to sum up what the men in Baixiu Village think about extramarital sex. To them, women's bodies are objects of pleasure and sexual satisfaction serving the sole purpose of men's sensual gratification. In their eyes, this sexual relationship becomes *the* relationship between the sexes. Gradually, the relationship is rationalized from the angle of the traditional father–son dyad. We saw in the last chapter that elderly and young men alike, people in the majority social stratum, and principals and teachers in schools seldom reflect upon such sexual questions from a moral angle. On the other hand, they are not without moral opinions on such matters. For instance, if a man is fined or imprisoned for having sex with a prostitute, he will feel a deep sense of shame. But this shamefulness is only a result of the censure imposed by traditional social morality; it is not an expression of regret toward the wife or children. Obviously, such an attitude toward women seriously affects the relationship among family members, and will in turn affect relationships with other people.

One may wonder about the woman's point of view. In this chapter, we shall describe how women feel about the sex phenomena in Baixiu Village. Three groups of women of different social positions are involved. The first group consists of sing-along girls in karaoke nightclubs, the second group is female factory workers from other provinces, and the third is local women.

How Sing-along Girls Feel

It was an evening in early spring; the year was 1995. Zhichao drove me on his motorcycle to a karaoke nightclub at a corner on Baixiu Village Road, where I hoped to find a suitable sing-along girl to interview. It was raining that night, in the air a whiff of winter cold still lingered. I had not dressed sufficiently and with rain beating down on me I felt the cold intensely. I had been running around between Qingyang Town and Baixiu Village these days, and the interviews had not gone the way I had hoped. I felt run down, in both mind and body.

We went into the nightclub and sat close to the dance floor. The table was a small round one, embraced by four low-backed chairs. We ordered drinks and took our time to survey the decor of the environment. The hall was about 400 square meters, rectangular, with a 150 square-meter dance floor in the center. At the front part of the dance floor was a slightly elevated stage for performers. On two sides of the floor were two rows of tables, behind them a row of two-seater carrels. The private rooms were on the other side of the main hall.

The attendant told us that the minimum charge was 25 yuan per person. A two-seater carrel was to be taken whole, with a minimum charge of 125 yuan. Private rooms ranged from 60 to 100 yuan per hour, depending on size; food and drink were not included in the prices.

The place was quiet, being almost empty. Only six or seven tables, or about one-eighth of the full capacity, were occupied. This was supposedly the largest nightclub in the vicinity, open for business only since the middle of last year.

I sat down with Zhichao, trying to figure out how to get a sing-along girl who had been in the business for some time for my interview. A deeper understanding of the job can come only with time. I knew from experience that nightclub attendants usually cannot tell which of the sing-along girls is the most experienced, and normally they do not entertain the patrons' requests. Moreover, patrons in general prefer newly arrived girls.

We sat there a long while. Still no other customers came. Zhichao joked that we looked like bouncers in the nightclub. Just then, three customers walked in—two men and one woman. They were from the same village as Zhichao. Zhichao said that one of them was a regular here and suggested that he would be the best person to introduce me to the right sing-along girl.

A little while later, the attendant led a sing-along girl to our table; she was about 160 centimeters tall, thin in build, and not particularly pretty. However, there was a sharpness about her eyes. Neither her dress nor her attitude marked her as a person in this trade.

She sat down, and struck up a conversation with Zhichao. She was from Julin in Guangxi Province, so she could speak Cantonese with Zhichao. Zhichao's Putonghua was weak. That might be why his friend suggested a Cantonese-speaking lady to attend to us.

The conversation between Zhichao and the woman was sporadic. Her name was Fang Ling, age twenty-two. She had worked in the nightclub for just ten days or so. Before that, she had been a receptionist in the biggest nightclub in Hongshi (Red Rock) Town quite near Qingyang Town for half a year. I thought that although she had been a sing-along girl for only a short time, she had half a year's experience as a receptionist and had probably witnessed a fair number of scenes and had some relevant experience. I decided that she was a suitable subject.

Zhichao carried on the conversation with great effort. I guessed he was trying to help me obtain information. But neither seemed very interested. Later, Fang Ling suggested playing the dice game "Call My Bluff." It was a game of wit; the loser had to drink a glass of beer.

Then Zhichao's friends also joined in the "warfare," but they kept losing. With the game of "Call My Bluff" came a more convivial atmosphere. I thought that the time was ripe to make an appointment with Fang Ling to work on my interview. I did not immediately tell her that it was going to be an interview. I simply said I would like to see her again in the nightclub the next day at three o'clock. I would explain myself then.

At about three o'clock in the afternoon the next day, I went to the night-club again. Experience told me that in general sing-along girls did not take such promises to heart. I hoped she proved to be an exception. With such thoughts I went into the hall. There were two female receptionists there, standing in front of the counter where the cash register was. They asked me if I had come to have a meal. There was a Western-style restaurant next to the hall, which was empty. I told them that I had an appointment with Fang Ling. The name did not seem to ring a bell with them. After some fuss, they made a call to the dorm upstairs to ask for Fang Ling. The answer was that Fang Ling had gone to Hongshi Town to see her sister go off to their village. On hearing that, at first I thought I should go. On second thought, Fang Ling had told me yesterday about her sister going home. Maybe she hadn't come back on time. I was there anyway, why not wait a little?

I sat down next to the counter and was served a cup of hot tea. There were no customers around, only a few receptionists and attendants, and one or two security guards. I had the feeling that they seemed to be looking at me with curiosity; few people would come to the nightclub at this hour of the day, let alone come unaccompanied. Fortunately, after less than ten minutes, Fang Ling came in hurriedly and said she was relieved to see me still waiting for her. I smiled in relief too, and asked for a private karaoke room, hoping to be able to conduct an interview.

The area of a private karaoke room was about 15 square meters, oblong in shape. On one side was a long sofa. On the other was a television set and equipment for playing videos; it was quite a good audio-visual setup. At the head of the room was a small open space. Installed in the ceiling was a set of revolving multicolored disco lights. The space and the lights might have been arranged to form a small dance floor for the patrons.

We each ordered our drinks. Fang Ling opened a thick album full of song titles and asked me which song I would like her to sing. I said it did not matter, she could sing whichever she pleased. She asked for my opinion before she sang each song. I just sat there listening to her.

Experience had taught me that I should not mention the topic of an interview immediately. First, there was the social condition in China. Second, because of the nature of the profession, sing-along girls were wary of such interviews. In fact, the great majority of the ladies had never come across, or even heard of, such a thing as an interview. And even if some agreed to answer questions, what they said was often neither complete nor true. So, to have a good interview, one had to gain the trust of the lady to a certain extent and become friendlier with her.

For about an hour, she kept singing. We occasionally had brief exchanges. I thought time was getting on, and so I told her who I was and my intentions. She was not too surprised, which I did not expect. On the contrary, she showed interest. This is how my first interview with her began. Subsequently, in different places, I conducted several in-depth interviews.

Fang Ling was a native of Julin, Guangxi Province. She completed junior secondary school and can speak fluent Cantonese. Julin was a small village. It is inhabited by the Fangs, who are Hakka and had migrated from Guangdong because of floods several decades ago. Fang Ling was the eldest daughter. She had a nineteen-year-old sister, the one who worked in a factory in Hongshi Town. Next came two brothers, one in senior secondary one and the other in junior secondary one. The family consisted of six people. Her father's leg had been injured in a traffic accident recently, and he could no longer walk properly.

"Farming was awfully hard work. But my parents only asked us to help when it was really busy, for example when transplanting rice seedlings," Fang Ling reminisced about her village life. "We had a large lake in the village, so irrigation was not a problem. The land was very fertile, and relatively speaking, ours was a rich village. We had rice in our three daily meals, but there was no spare cash. Father took what was left of the rice to sell and could get about 3,000 yuan a year, half of which went to buying fertilizers.

"About three years ago, two young women from the village went out to Hongshi Town to work. They came back one year later. Since then, many villagers have followed their example. Now, men and women under thirty come to Guangdong to work."

Fang Ling had gone to Hongshi Town to become a factory worker a little over a year ago. Then she transferred to a clerical job. The work was hard and exhausting, so she quit and went to the largest nightclub in Hongshi Town to become a receptionist. After half a year she changed jobs and became a sing-along girl in the present karaoke nightclub. I asked her why she had switched from a clerk to a receptionist.

Fang Ling answered, "The month as a factory worker was killing me. Later the boyfriend of a friend got me the clerical job, with a monthly salary of 600 yuan. Then I became a receptionist. Because of that, I often went for karaoke with friends for enjoyment. Gradually my thinking changed. At the same time, I saw some friends of mine becoming the bought second wife of Hong Kong merchants. They live a life of comfort. They have a monthly income of over 10,000 yuan. They are provided with all sorts of electrical equipment, and often receive presents from their male friends. They go shopping every day. They are dressed smartly and do not look like those 'ladies' at all. They often say to me, why not just become someone's 'second wife' and be done with it. Sometimes I feel the same way. If you ask them whether they love their boyfriend or not, they all do, very much. They are happy. Of course, their boyfriends also love them, otherwise they would not send them all those presents."

Fang Ling's views and the change she was undergoing were very common here. The other sing-along girls that we interviewed, especially those who were factory workers, were largely affected by their peers. They also regarded being the lover of a Hong Kong or Taiwan merchant as the best possible destiny. But what Fang Ling meant by "love" is not romantic love between a man and a woman—she meant only the material provision offered by the boyfriend. And the women's attitude toward the men is in most cases not really "love." Perhaps it could count as "liking."

Like all the other sing-along girls, Fang Ling did not like the local people. She said, "I don't like the locals. They look down on us outsiders who come here to work. Sometimes when we meet them on the street, they give us a telling-off.

"The locals often don't sing when they come to the nightclub. They like to see porn videos. They come for the purpose of enjoying the ladies. We don't like the locals. Taiwanese are all right, but chauvinistic and often force the ladies to drink. Hong Kong men are the best. They are the most polite and would not go wild in a private room. If they want to make love to a lady, they take her to a hotel or the dorm of a factory. A local person would sometimes do it in a private karaoke room if no third person was present.

"I was given a dressing-down by the manager twice for offending customers. In both cases, it was because they were bad. One time, a local customer pushed me down onto a sofa in front of other customers and ladies, and started groping my body. I pushed him away and walked out of the room and would not go back. And then the manager gave me a dressing-down," Fang Ling said, anger audible in her voice.

"If a customer tried to caress me, I would think of some ways to divert his attention, for example, ask him to sing a song or have a drink. I don't go out to spend the night with customers. I don't want that sort of money. But, of course, if someone offers me a good price, I will. Or if he gives me 10,000 yuan a month, I would not mind being 'bought' by him."

None of the sing-along girls we interviewed liked the local people. And generally speaking, the local people looked down on Mandarin-speakers from other provinces. They were especially contemptuous of sing-along girls. On the other hand, many ladies working in karaoke nightclubs seemed quite willing to marry a local man and settle down. This is similar to the attitude of women factory workers, who will be described in the next section. In fact, although the locals were said to be rude to ladies, according to our observations, men from other provinces were even ruder. Sing-along girls, especially, were often abused by out-of-province security guards as well.

This picture reveals how the prosperity of the Pearl River Delta is attracting the sing-along girls and women workers of the less developed regions. Perhaps another reason is that in the Pearl River Delta, these women are far away from home, and this accords them greater freedom in determining their own lives. We can see this freedom in the words of Fang Ling.

"To choose between my home village and this place, I would rather stay here. By comparison, life in my village is too harsh, and it is just not as prosperous or eventful. What's more, you earn more money here. The fact is that many parents in my home village would like their children to go to work in Guangdong. Many of the new houses built in the village are paid for by men and women who work outside and send money back. Take me as an example. I often send money and stuff home. Sometimes an article of clothing may not be needed anymore, and I send it to my mother. Other people seeing that are envious of her. They say she has a daughter that shines. Their children work in factories and do not send home so many things."

Fang Ling seemed proud of her earning ability as a sing-along girl and a receptionist, but at the same time, she cannot rid herself of the sense of shame associated with this profession.

"We try our best not to let our families know that we are 'ladies' or 'bought second wives.' I know of a girl from our village who became a 'bought second wife.' Her situation was known to her family. At first, they were scornful of her. Then she got married to a forty-something Hongkonger who builds roads in China. She now lives in Shenzhen and her husband comes home every week from Hong Kong to see her. Her status in the village is now much higher, and villagers try to get near her. It is a great honor to be married to a Hongkonger."

Interviewing sing-along girls is not easy, and was especially difficult in the mid-1990s. The nature of the job is such that a lady will not easily trust anyone, especially a man. Even for those in professions where sex is traded for money, as soon as one brings up man–woman relations or sexual matters, Chinese women are very conservative. Because of the social taboo associated with sexual matters for thousands of years, women do not say what they really think where sex is concerned. They express only what social standards allow them to say. The fact is, even if they really speak their minds, what they say is often inconsistent. To a certain extent, their viewpoints truly reflect how they feel. They are far away from their home village; they ply the sex trade, selling their bodies. In addition, Chinese girls, particularly those who grow up in a village, tend to marry young. So those women who have been in these trades for a while start to worry about their future, especially their marriage prospects.

Moreover, under the present circumstances in China, women engaged in such jobs are all the more afraid of interference from political and military *danwei* (work units) personnel. As their knowledge and vision are more restricted, it makes in-depth interviews that much more difficult. So, although Fang Ling had been on the job for only a short time, her candid talk and her observations from when she was a receptionist were all the more valuable. In fact, Fang Ling could articulate her feelings more clearly than other interviewees. By the same token, her perceptions of her own future was clearer, but that much more hopeless.

After my last interview with her, we had lunch in a restaurant near Baixiu Village. Following the meal, I accompanied her home to her dorm. It was a sunny afternoon, but the spring wind weakened the sun's warmth. It took about ten minutes to go from the restaurant along Qingyang Main Road to her living place. The sun was at our back, its rays intermittently pouring between tree branches onto Fang Ling, casting a long slender shadow of her on the ground.

"Pity, the days when I was a receptionist were my happiest days," Fang Ling suddenly sighed. "I got on well with my workers and manager, and we looked after each other. I am not happy over here. When I got here last month, there was a raid on the sex trade by the police. I did not get to serve customers every day. From seven in the evening to midnight, I would sit there. What an awful time it is when nobody wants you!

"But when you are called to a table, people take liberty with you—they grope and they touch, especially the local people," she continued. "Being a sing-along girl is an empty life; you are wasting precious time. I was happiest as a receptionist."

"You could go back to the factory. You were making 600 yuan. It is not a small sum, don't you think?" I said.

"Yes, it is not small. But I am used to spending money now. I don't want to do such hard work in a factory. Yes, it is not easy to turn back. Maybe I will do as my friends do, get 'bought' by someone." She seemed to be muttering to herself, but her tone sounded quite serious.

To tell the truth, Fang Ling was not beautiful. She did not have the qualities that catch the eye of a Taiwan or Hong Kong merchant looking to "buy." She probably knew this. That might be one of the reasons she felt perplexed and hopeless about her future. As explained before, this shows what the "ideal end" for women in these jobs is. More generally, being a second wife is what many women from other provinces hope for after working in Baixiu Village for a period of time.

The Perspective of the Women Workers from Other Provinces

Zhang Ting, age twenty-one, comes from Guangxi. She has primary school education. She is 165 centimeters in height, slender in build, with long hair. Not particularly beautiful, she is nevertheless rather charming. She has two brothers and one elder sister. Her sister is the second child and she the third.

Zhang Ting came to Baixiu Village to work in the late 1980s. At that time, her second brother did not agree to her leaving home. Her sister had left home and worked for two months. When her brother wrote to demand the sister's return, she obeyed. But not Zhang Ting. Her temperament is such that she *will* do the things people forbid her to. She said that she has something of a "rebellious" character.

It is Zhang Ting's view that pretty, educated, and eloquent girls will get better jobs, pretty ones all the more so. There was a girl in her village who was beautiful. As soon as she got into a factory, she was singled out by the manager and appointed to a job in the office. This can happen even to the not-so-well educated. Zhang Ting considers herself neither pretty nor cultivated, so she has stuck with her job. She's been working for a number of years now, still making only three to four hundred yuan a month.

Zhang Ting declared, "I have been out these many years. What have I got? Nothing. And the ideal man? None." She said her standards were not high, but she could not find the right person because people are too complicated and there are too many bad people around. "They are never trustworthy—these people from out of my own village."

I asked her how she got on with the local people. She replied, "When I first got here, Baixiu Village was not so prosperous. The people were hospitable. However, as rural reform progressed, the villagers became richer, and they also became colder. Everyone is selfish now, thinking of themselves. They have become terrifying to people from out of the province, thinking that since they have now got rich, they can do anything they like. Just a few days ago, a local person driving a motorcycle crashed into an outsider riding a bicycle. The outsider had to pay 200 yuan, or else suffer a beating. This sort of thing happens all the time. The people here have become crazy."

Zhang Ting's biggest wish now is to marry someone well-to-do. She said, "Everyone is of the opinion that once you marry someone well-to-do, then you don't have to work so hard. I continue to hope to get a job in a better company. My wages are only 400 yuan or so—it's too little. Seven hundred to 800 yuan will be about okay. Now the living conditions are poor. Eight people crowd in a room; that's too many, and the bathroom is shabby. The food I get is awful. Eating badly will lead to sickness. I don't like the food in the factory; I often eat instant noodles."

"Why don't you change to another factory?" I asked her.

She answered, "Not that easy. If you don't have friends, you don't know what kind of factory it is. My cousin changed factories three times this year; now she is without a job. She fools around every day. Now it's not so easy to get a job in a factory. People expect you to be educated. Some companies look for good-looking women, and if you look pretty, they will take you even if you've only had a little schooling. I have neither a pretty face nor a good education; I'm done for."

I said, "A pretty face gets you a job more easily. Doesn't that make a pretty face the envy of everyone?"

"It does, and I am no exception," she couldn't help breaking into a smile as she came to the second part of the answer.

"Do you approve of pretty girls becoming sing-along girls?"

She replied, "It's all right for those who want to become one. For those who are tricked into becoming one, their beauty is their downfall, you might say. Even for those who do it of their own will, they suffer the consequences later, not now. When they get married, if their husband finds out that they were once a sing-along girl, there would be a squabble and then divorce."

Zhang Ting's view is typical of the group of women workers from out of the province who are not well educated and not that attractive. Because they do not have any special qualities to speak of, they can only move about in low-paying positions, without any hope of promotion. The rural economic reform has brought opportunities: they can earn more money and see the world. But leaving home means raising hurdles in the pathway of marriage. These problems are also evident in two girls from the province of Hubei, Li Mei and Wang Qingmei. Both are well educated and can express their feelings more adequately.

Li Mei is twenty-two years old, a secondary school graduate. She has a sister and a brother. Once a teacher in her village school, her ambition is to enroll in a music academy. She came to Baixiu Village a year or so ago, and because of her higher educational qualifications, she got a job as an office worker in a factory. Wang Qingmei comes from the same village as Li Mei; she is similar to Li in background and education. She has a sister and two brothers. She also came to Baixiu Village and joined a factory about a year ago; now she is in charge of one production line.

Li Mei is a little on the plump side, 160 centimeters tall. Her skin is ivory, her hair cut short, and she wears black-rimmed spectacles. Her face can be said to be fair. Wang Qingmei is smaller in stature, about 155 centimeters tall, long-haired with an unremarkable face. They were together throughout almost all our interviews. I had sporadic talks with them as an acquaintance for two years.

The first interview took place at the research center. Among the many topics covered, one was my query as to how they viewed karaoke nightclub girls. Wang Qingmei thought that if the job is simply to sing along with the client, and get tips for it, then it is all right provided the woman does not have to give herself up. Li Mei agreed with Wang Qingmei, and said that as long as one does not sell one's body, then it is okay. Li Mei elaborated what she meant and said that the general public definitely looked down on karaoke girls, but if a woman had self-respect, then she could hold her head high no matter what other people said.

Li Mei added, "Karaoke girls can put on a happy smile to please a customer whenever required. That is some trick. Sometimes a customer will take advantage of her, pinch her. Even then, that is all part of the equation, and a girl is psychologically prepared for it. The same things happen in the factory. The weaving department has a new worker, a divorced fellow. He likes to take a light swipe at the women workers' busts or bottoms. Most women workers did not take offense; they just laughed. You see, in the case of the workers, they are not psychologically prepared, yet they laugh spontaneously. I think they are just an inch away from being profligate. To my mind, the karaoke girls are better than they are."

This reply surprised me. It seems that as long as a sing-along girl does not sell her body, then everything else is acceptable. I followed up, "Then why don't the women workers take up the job of being a karaoke girl?"

Wang Qingmei glared at me. "You've got to have the right stuff. Not everyone can do it."

"Yes. You've got to look pretty and put on a smile readily," Li Mei added.

This shocked me a little, as if they thought that being a karaoke girl was something to be proud of, and they could not become one even if they wanted to. It was something to be admired, not despised.

I pursued the topic: "Do you think that they feel pain privately because they have taken a job that goes against the traditional moral teaching they have been given since very young?"

Wang Qingmei said, "Let me tell you. If a client is generous in rewarding her and nice to her, then she will not feel pain."

Li Mei carried on in the same vein, "What she meant is that plenty of money will smother the pain, because money's what they want in the first place."

I replied, "That surely must leave a dark shadow in their heart."

"People nowadays are broadminded. Who cares that much? She can't very well go and commit suicide." It was Wang speaking this time, and after a pause, she continued, "The prostitutes in history have high standings. Look at Li Shi Shi in the Song dynasty. The last emperor of Song really loved and prized her."

To some extent, the views of Li Mei and Wang Qingmei are representative of those of many women workers. It could be that they carried a grudge against those who got promoted on the strength of their looks, and so by way of contrast praised the sing-along girls.

I asked them to compare their lives in the village and their lives here.

"Before, in the village, I got a 100 and more yuan a month teaching. That was decent pay. The circle of friends and relatives was small, and life was not exciting. All I wanted was to find a like-minded boyfriend, get married, and live a peaceful, happy life. When I got here, I saw that the life people lived was much richer than that in the village. In my home, we still use hay and coal as fuel. The wok is all covered with soot on the outside. Here everyone uses electric cookers and burns natural gas. In my home, the floor is just smeared with concrete. But here, the houses are beautiful, the floors paved with ceramic tiles. The everyday tools available here are no comparison. Our village just doesn't have them. The things people use here are modern. Once I saw how things were here, I did not want to go back to the village and live the plain life and eat rice out of a blackened wok. Even the criterion for finding a boyfriend has changed. I used to put more weight on affection for each other, now it is more money." Li Mei was frank.

It was several months before I met Wang Qingmei and Li Mei again. I met Zhichao when I came to Baixiu Village on a visit, so I rode on his motorcycle to go to the factory where Wang and Li worked. In the evening, I had a meal with them in the small restaurant beside their factory.

Wang Qingmei was her old self; Li Mei had changed somewhat—her hair was permed and her face was a little haggard. We updated each other as we ate.

"It seems that all the cinemas are showing pornographic films," I remarked inquiringly.

"They all do after nine o'clock at night. Now a few more cinemas have started business. They all compete with each other on the kind of pornographic films they show," Wang Qingmei said.

"Are the cinemas full?" I asked.

"Yes, the workers do not have many places to go to," Wang Qingmei answered.

"Do girls go too?" I wondered.

Li Mei said, "Of course. There are six girls in our dormitory. Four of them have become addicted to it. They never go to see a seven o'clock show, always the nine o'clock show. Sometimes they will see three shows in a row. When they get back to the dormitory, they will talk about the films. These become the topics for jokes."

"Have you seen these films?"

Wang Qingmei giggled, pointing to Li Mei, "She took me to see it once."

"There was not much there that time, only a few scenes showing some exposed parts of the body," Li Mei said.

Wang Qingmei continued, "I think if it happens naturally, then it is acceptable. What is awful and cruel are those stories about a woman being kidnapped and forced to do things."

Our conversation drifted to the topic of the local people's marriages.

I said, "Since the men in Baixiu Village go for nightclub ladies, won't the local girls find it hard to get the right husband?"

Wang Qingmei replied, "The women here are all old. Does Wang Guizhen have a boyfriend?" They worked in the same factory as Wang Guizhen.

I shook my head, "I don't know."

Wang Qingmei continued, "No. The local girls are none too pretty. Who wants them? Besides, you have to be serious with local girls. Not the outsiders. Once you have what you want, you can just go."

I asked, "If married men go to see nightclub girls, will it not affect the stability of the family?"

Li Mei said, "The wife has to be tolerant, with one eye shut and one eye open. As long as he does not divorce her and there is money for keeping the family, then it is all right."

"Is this true in your home village?" I asked.

Li Mei interjected, "No! There would be a big hullabaloo if it happened in the home village."

Wang Qingmei explained, "Actually, in the home village, if a man goes out trading, their family is not really stable. But then, when the man goes out to do business, the wife takes care of everything in the family. She also has land. She grows the subsistence crop and the vegetables herself. Over here, there is no land women can attend to. They also go out to work, and cannot take much care of the home. They do not earn much, yet the cost of living is high. The pay just does not cover the expenses. As a result, women are also dependent upon men."

Then Li Mei spoke. "The truth is, both in the home village and here, there is male chauvinism, only it is more apparent here. Once, several of us women workers, the supervisor, the factory manager, and their wives went to a karaoke nightclub. The manager did not stay together with his wife to have fun, or with us, but he asked for a sing-along girl. His wife did not seem to mind at all; she took it as if it was the most natural thing in the world. If this happened in our home village, there would be great disturbances. Men in the home village will only do this behind the wife's back, not right in front of her. They would not embarrass her like that."

In the evening, I went to the appointed time and place to wait for Li Mei. That night we were to meet alone. She appeared, an outstanding figure among the crowd, with a pale yellow dress and a look that was a little wild. She was walking in my direction, but she did not see me. I called out "Oi" and then she said, "Oh, you are here."

We walked together. Along the way, people who passed by turned their heads toward Li Mei but she ignored them. I could tell, though, from her stiffly straight back and neck, and the expression on her face, that she was proud of her unusual outfit and her attractiveness. Her dress was beautiful, though a little frayed at the neckline and the sleeves.

Li Mei was supposed to work overtime that night, but she said that getting leave was easy. Walking along the road, she told me, "Tonight our supervisor won't work overtime either. He has to see his girlfriend off. She was fired by the factory manager."

I thought of what Wang Qingmei had told me. She had said half-jokingly that Li Mei had her eye on her supervisor. Li Mei blushed then and said, "My supervisor's girlfriend is too uneducated. When they go out, they have nothing to talk about. That's why he prefers talking to me. We can have good conversations."

I asked directly, "What is your relationship with the supervisor now?" Li Mei blushed again and did not answer.

Later, Li Mei told me about her promotion. She was excited. "Now I have more than 700 yuan a month. After the promotion I shall have almost 900 yuan, more than Wang Qingmei!"

I asked her how she spent her money. She told me that since coming to

Baixiu, she had sent home 1,000 yuan. Almost all of what was left of her savings she has lent to people. I said, "Now that you have more income, you can afford some dresses."

Li Mei agreed. "Yes. I think when I buy a dress, it will be one that can be worn on special occasions, one that looks grand. Up to now, I have not bought really expensive clothes. The clothes here are awfully expensive. Take the dress I am wearing. It would cost 70 yuan in my home village. The other day, when I was in Qingyang Town, I saw one exactly like it for 200 yuan!" Li Mei paused for a little while, then continued, talking more to herself than to me, "I don't want to buy really expensive clothes. Actually we have to wear work uniforms every day at work. Holidays are few. There are not many occasions when nice clothes are appropriate. Besides, if you are too well dressed and spend a lot on clothes, then people think that you are in that kind of job."

I wanted to find out about the relationship between the women workers and the local men or men from Hong Kong and Taiwan. I asked, "Are there some women workers who are 'under the protection' of men?"

She lowered her voice, "There are two in our factory. One is the manager's mistress, one is the supervisor's mistress." I said, "Your supervisor?" "No," she said, "Another supervisor. What my supervisor has is a girlfriend. We make a distinction like this. If a man is married, has children, is rather old, then we call the woman his mistress. If he is not yet married, and has only one woman friend, then we call her his girlfriend."

I asked, "Is this widely known?"

She answered, "Yes, everyone knows! Everyone is used to it. When the manager goes around with a lady who is younger than his daughter, we aren't surprised. In fact, it seems quite the natural thing." Then, Li Mei changed to an incredulous tone and said, "You know, there is a foreman who is going out with the manager's mistress. He knows full well she is the manager's woman. Well, it's strange, isn't it? Men usually fear a woman's fickle! Maybe it's because she is in charge of the personnel department."

"Doesn't the manager mind?"

Li Mei threw me a stare. Rolling her eyes, she said, "What's there for him to mind? As long as she goes around with him and plays with him, what else does he want?" Li Mei was either asking me, or asking herself, or neither. I felt that she was a little numb. I found her character to be something of a contradiction. My opinion is that Li Mei is impulsive, yet her education makes her think deeply about the consequences of her actions. She has knowledge, therefore a stronger sense of morality; however, faced with money and power, she cannot help evincing a strong desire, a lust, for them. She is always caught on the horns of the dilemma. She is now in the executive rank. Compared to workers, she is faring better. She must have a certain sense of satisfaction. Her way up, it seems to me, is to improve her social standing and income through promotions. Unless something drastic happens, she will not go the way of the mistress. After a while, I asked her, "How do you see such women?"

Her brows knit, Li Mei said, "I don't say anything to them to their face, but my heart feels repulsion." After a while she added, talking to herself, "I can't understand why they are willing to become mistresses." She seemed to have fallen into another contradiction.

I said, "For money of course."

She stared at me. It was as if I had been too raw in my reply, and it threw her off a bit. After a pause, she said, her brows knit again, "But this becomes known here by people from the same home village. How can she face the other villagers when she goes back home?"

I then asked Li Mei, "Is it true the local women do not like the women workers from other provinces?"

Li Mei was agitated. She breathed heavily, "Indeed, they look down on us. We really work as well as they do, even better."

I said, "Is it because their husbands fool around with outside women workers?"

Li Mei seemed to suddenly see the light. "Oh, is that how you look at it? I have never heard them say that's the reason."

It was more than half a year before I went to Baixiu Village again. I reached Baixiu Village at noon with my other research colleagues. I phoned Wang Qingmei and Li Mei, making an appointment to have dinner with them; then I went to visit other friends.

Baixiu Village had changed a lot in the span of more than half a year. The marketplace, which used to be full of hustle and bustle, was now somewhat quiet. When there were factories in the village, during the noon break, people from the factory dormitories and those just passing by crowded the road, coming and going. The small shops lining the two sides of the road were filled with workers buying daily necessities and miscellaneous goods and eating lunch. In the past, at this hour, I might sit in Ah Xing's photography shop at the street corner, chatting with Ah Xing's employees and watching the people pass by, saying hello to those I knew. I would wait there for friends who had made an appointment to meet me, then go out for a meal, for a good talk. This time, when we reached the street corner, we saw Ah Xing's shop firmly shut with a steel plate. The road in front of the shop was deserted. People told us that since the factory zone had been built on the other side of the village, most factories had moved there.

At six o'clock in the evening, I waited in front of the dormitory for Wang Qingmei and Li Mei. As soon as we met, Li Mei blurted out, "Do you know, I am now in charge of one production line!"

"So that makes you a line leader," I said.

"Yes," she replied in her sweet and high-pitched voice. She looked both proud and somewhat uneasy. During the meal, I asked them how the factory was and what their work was like.

"I am very busy; she is not." It was Li Mei again. "But, being busy is good; otherwise, you don't know what to do with yourself." She told us that when she returned to the factory from her home village after the spring festival, business was slow in the factory, and there was plenty of free time during the week. Daily wages for the new workers had risen from eight yuan to nine and a half. Overtime pay was still 3 yuan an evening. After a little more casual conversation, I asked Li Mei, "Are there still that many karaoke nightclubs in the village? That many girls?"

Li Mei gave me a staring look, and at the same time said slowly with pouted lips, "I don't know." I felt it embarrassing to press the point, so I talked about something else.

After dinner, Wang Qingmei went back to the factory to do overtime. Li Mei stayed to keep me company a little longer. We stood at the door of the dormitory. Some women workers went past us. They were dressed nicely. From my observation, even the "nice" clothing worn by some workers window shopping on the street on nonworking days was cheap, quite recognizably so. However, the dresses worn by these women workers were of considerably high quality. Both the cut and the materials were above average. I talked with Li Mei some more, eager to find out more about the relationships between the women workers and their male bosses. Li Mei's stare instantly warned me. I dared not ask directly. I changed to a more tentative tone, "Are there many in the factory who are mistresses of other men?" No sooner were the words out of my mouth than I felt they were too blunt. My worries were not well founded, though, for she answered immediately, "Too many of them." Her tone was unruffled, as if she was talking about something that should cause no surprise. A pause, then she continued, "I am used to seeing such things now; I don't care, as long as I do not do it myself."

I changed the topic and asked her about her plans. Will she continue like this? Does she still want to go to an academy of music? I said that her rapid promotion was not easy.

She was gratified by my compliment. She said, "I shall soon be promoted to second rank, then I'll be more senior than Wang Qingmei. Those locals who are illiterate have become line leaders. My department has nine line leaders, six of them local, two of them from the same home village as the supervisor. I am the only one who hasn't any *guanxi*. Unless you are very competent, you can't get further promotion without *guanxi*. But . . . I'll probably stay put. When I went home during the spring festival, I was so unused to the place. Whether I stayed in my home, or went out in the village, it was dark and dismal. I stayed for just a few days, and I thought of coming back here all the time." She paused, then said, "Now my monthly wages are more than 700 yuan, more than 900 yuan with the bonus. I'm better off than some, worse off than others. When I have saved some more, then I'll consider the question of further education. Maybe I'll go home and do business. Those are the only two choices."

As we talked, a motorcycle zoomed past us. Right after that, another one drove past. Li Mei pointed to the faint shapes in the distance and said, "They are two sisters, both married to local men."

"Oh?" I was somewhat surprised. "What a coincidence! Both married to local men!"

"Yes, they came here in the early days. At that time, there were not so many restrictions about marrying locals." Her expression was one of envy.

I had hoped to have a meal with Zhang Ting the next day, but somebody told me that she had gone back home. I was momentarily surprised. But then, outside workers coming to work here belong "outside." The restriction of the household registration system, the exclusion by the locals, the traditional idea that men and women reaching adulthood should be married—all these made most of the women, especially those reaching the late marriageable years, decide finally to return to their home villages after working for a period of time outside. Zhichao later told me that Zhang Ting's parents had found a boyfriend for her, so homeward she went. Hearing that, I was somehow deeply touched. I remembered that when I had chatted with Zhang Ting some time ago, she kept saying that she was neither beautiful nor well educated. And after five or six years' laboring, she could make only 400 or 500 yuan a month. She kept comparing herself with those women who were pretty and well educated, and who got promoted easily. I thought Zhang Ting must have felt a sense of injustice and at the same time a sense of helplessness. Sometimes she would say to me, in earnest, "Getting married to a good husband is a woman's biggest blessing." At other times, she would say, jokingly, "I'll never get married." Zhang Ting returned home in the middle of last year. By now, she has been gone for half a year. Maybe she has made herself a daughter-in-law to a new family.

Activities of Sexual Indulgence in the Eyes of Local Women

I had just paid a visit to the female owner of a small shop right opposite the market in Baixiu Village. She was the wife of the eldest son of the Wang family, Rongzong. People called her Rongzong Sao (Aunt Rongzong). Her mother-in-law had passed away already. Her father-in-law was not living with her, but he often came to see his son, daughter-in-law, and grandson. During big festivals like Qing Ming, Mid-Autumn, Lunar New Year, and so on, Rongzong's home would become the focal point for familial activities. Rongzong Sao became very busy during these festivals, yet she was very enthusiastic. Ask anyone in the village who Rongzong Sao was and surely they could all give you the answer. Everybody knew her here, and whenever they mentioned her, they also mentioned her husband Rongzong.

Like other men in the village, backed by the lineage of Wang as well as his own knowledge and abilities, Rongzong was now the head of a relatively large factory. The couple was a good match. The husband was close to his forties, slim, and had his hair permed. The wife was over thirty, had a slightly pointed face with a short pig-tail tied up at the back. The two of them went through the process of courtship before they got married. Their married life seemed quite happy. During holidays, the husband would take the wife out for trips. During ordinary days, Rongzong Sao looked after a small shop that sold snacks. Their life overall seemed quite happy and satisfying.

In contrast with other shops nearby, Rongzong Sao's shop did not supply daily necessities like rice, oil, or salt. Nevertheless, there were always three to four people in her shop, most of them women. Apart from Rongzong Sao's aunt, there were also other local housewives and some women who came from other places and had married local people. They liked to sit on the steps in front of the shop and chat.

Rongzong Sao was very happy to meet me. After having talked about some of our recent activities, I approached her directly on the issue of the sex trade prospering in the village during recent years. Rongzong Sao seemed taken aback. She paused a while and said, "This has to do with the city. We don't have that problem here."

I said immediately, "Along the road entering the village there are a lot of karaoke bars and nightclubs. They provide girls to keep customers company and sing along with them, they also provide 'those' services."

Rongzong Sao did not show any special response. She looked glum. I continued, "I've heard that in Baixiu Village, a lot of female workers who came from other places have developed illicit relationships with some local men here."

Rongzong Sao still looked numb. She stared blankly in the same direction and said, "Those who are married would not get involved."

It seemed that Rongzong Sao wanted to change the subject. Possibly, either consciously or subconsciously, she wanted to evade the issue of prostitution and "buying a second wife" in the village. As such, I did not intend to continue the conversation with her. After a short while, I said goodbye to her and left for Dayan Sao's home.

At first the response of Rongzong Sao had seemed somewhat unexpected. Nevertheless, after more thought I viewed it as a "defensive" response, similar to that of some of the other village women. When faced with a hard reality that they could not change but felt sad and helpless about, they could only respond by adopting an almost oblivious attitude. Later on I found that Rongzong, her husband, actually did keep a mistress.

I arrived at Dayan Sao's place and got myself situated. As usual, she took out an apple from the refrigerator for me to eat. She was sewing her buttons when she said joyfully, "From this month onward I'll be doing some cleaning for the factory dormitory. I clean twice a day and every month I can earn 250 yuan."

Dayan Sao said that usually she would do her first cleaning from eight to nine o'clock in the morning, then she would do her second cleaning from noon to one o'clock in the afternoon. As far as I knew, every morning between eight and ten Dayan Sao used to ride her bicycle to her mother's place in the neighboring village so that she could chat and entertain herself. Therefore, the new job would affect her time at her mother's place. However, it seemed that she did not mind. She said, "When I can earn money and have a job, I won't feel so bored."

After a while, I tried to bring up the topic of the sex trade. At first I was a bit hesitant because I was afraid that she might react in the same way as Rongzong Sao. I was beating around the bush and asked her about those women who were involved in "that kind of business." At first Dayan Sao did not understand and she frowned. Later on, I made it more explicit, and she seemed to suddenly see the light. She straightened her back and leaned slightly backward. With her eyes opened even wider, she said to me in a loud voice, "Oh! You mean those 'whores'?"

Her response surprised me because she was so different from Rongzong Sao. I guess apart from personality difference, the reason why Rongzong Sao evaded this topic was because her husband "bought a mistress." As for Dayan Sao, her husband was far away in Hong Kong; therefore, sex issues in the village did not have a direct impact on her. Dayan Sao put aside her sewing and said to me seriously, "It's only about ten years ago that people started to visit prostitutes."

I asked Dayan Sao how many men in the village would likely visit prostitutes. She thought for a little while and said to me again, seriously, "Ninety men out of a hundred would visit prostitutes; among those at least fifty would 'buy a mistress.'

"Those prostitutes all come from places outside the village. Those who work in hair salons are certainly prostitutes. Apart from that, there are also women who work on a job in the daytime, and do 'that kind of business' in the nighttime." Dayan Sao kept on talking.

During the conversation I specifically wanted to explore how Dayan Sao viewed women who were in the sex trade.

"They just stick to you and won't let you go. What can you do?" She gradually showed some anxiety, frowned, and said, "Those women workers always hang around the men and won't let them go. These men include factory managers. Some are local people. Once they get their hands on you, they will stick to you and never let you go!" Dayan Sao put on a long face and showed her contempt.

During the later part of the conversation, Dayan Sao repeatedly said similar things. She did not have any further views on the topic. I thought this was because she did not have much education and therefore could not express herself clearly. It seemed that Wang Guizhen might have a better picture about these issues and in addition would be better able to express herself.

Although Wang Guizhen (Ah Zhen) was relatively younger and had a higher level of education, I still hesitated to bring up the topic directly due to the experience I had had with Rongzong Sao. At the start, I only asked Wang Guizhen some factual questions. I mainly asked her if a divorced woman had to stay in the village of her husband in order to get her share of money from the work brigade every year. I asked that because I thought in order to avoid calling up the pain of a broken marriage, a divorced woman normally would not want to stay in her ex-husband's village. Also, by staying away she could avoid being treated with contempt by the family of the husband. According to Wang Guizhen, the situation was that while the divorced women themselves would move back to their mothers' places, their registered residence would remain with their ex-husbands' villages. This way, they could on one hand get their share of money from their ex-husbands' teams and, on the other hand, they could get support from their own relatives. When opportunities came they would look for another husband.

I gradually turned to the topic of the sex trade in Baixiu Village. I told Wang Guizhen that some people said ninety men out of a hundred would visit prostitutes while at least fifty of them would contract a mistress. Wang Guizhen thought for a while and said to me with contempt, "I think in Baixiu Village eighty to ninety men out of a hundred would visit prostitutes. Among them at least seventy would buy a mistress."

I asked her when such a change had occurred. Her expression was rather deadpan. Nevertheless, she did answer my question. "It was about ten years ago. Those women who came from the north always wanted to get hold of those men in the village." Suddenly Wang Guizhen stopped. I was a little bit surprised, but I went on asking, "In your factory you have a lot of female workers who came from places outside the village. Do they do that kind of business?"

Wang Guizhen thought for a while, and then said to me seriously, "In the office, if I happen to know that you are somebody's mistress, I would consider that as your private affair and I would not bother you. I would still make friends with you. I am like that. But Zhichao Sao is quite different. She doesn't like them. Not only does she not interact with them, she also tells other people not to have anything to do with them. But I won't do that."

When Wang Guizhen said this, she was a bit nonchalant. Although she said that she would not be bothered about other people's private lives, and she appeared to be open, with a "business is business" attitude, she was obviously different than before. Between her facial expression and what she actually said, there were signs of discrepancy. I asked her further how she felt about the development of the sex trade in the past decade.

"I find it very repulsive!" She said emotionlessly, still looking nonchalant.

Having talked about this and that, we touched on the issue of marriage. Wang Guizhen said, "Some of my colleagues advised me to get married soon, but I just wonder how I can do that. How can I get married to somebody I don't like?" She paused a while and then continued, "When it comes to marriage, I find it very difficult. Sometimes I wonder if I can ever get married! In the first place the other party may not have the qualities I want. I want him to be stronger than me. You know I am relatively individualistic. I have my own views. I don't want him to obey everything I say. I hope he can think more independently than I do. You know, if I can't get along with my husband after marriage, it would be hard for me to take. Therefore, unless I really like him a lot, I won't get married. If anything goes wrong after marriage, it would be even worse than if you had never gotten married! To be honest, in the end, a husband is not reliable. The only solution is to give birth to a son. Only a son really belongs to you."

Having talked with Wang Guizhen, I had the impression that she was not eager to discuss the sex trade in Baixiu Village. She quite consciously showed that she held a liberal attitude toward it. She used privacy as the excuse for not making a moral judgment on such behavior. During the conversation, Wang Guizhen wanted to point out how the influx of outside female workers had created the phenomenon of local men visiting prostitutes and keeping mistresses, and how this was affecting the local women. Apparently, the women in the village could only passively and helplessly accept this reality and make their own adjustment in regard to matters of marriage.

In view of her attitude, it seems that the way Wang Guizhen looked at the issue of prostitution and keeping a mistress was not as open as what she said during the interview. As a matter of fact, she was like other local women in the sense that she was also a victim of the sex trade. Even if she could not "get herself a husband," with her high school qualification she was well equipped to search for other areas of development. However, other local women like Dayan Sao, who had received little education and did not have any particular work skills, could only rely on getting married as their means of livelihood. When faced with the competition from outside women who had better looks, a lot of local women in the end could only get married to men much older than themselves. Most of them were Baixiu Village people who had left for Hong Kong to make a living in earlier years.

What we can see here is that familism, or the "father–son dyad" kind of human relationship, is still deeply affecting the relationship between the sexes in Baixiu Village. Certainly we understand that the traditional mode of "senior superior, junior inferior" human relationship has been greatly reduced in today's China. Not many people will absolutely obey the will of their elders anymore. Nevertheless, in the relationship between the sexes, we can still see the "male superior, female inferior" shadow, which is inherent in a patriarchal culture.

It is good that in comparison with traditional China in the past, the women in Baixiu Village, be they outside female workers or local women, all have a lifestyle and outlook different from their predecessors'. They obviously have more control over their own lives and are able to make choices. That is a result of the market economy. On the other hand, in areas other than economic development like marriage or issues related to sex, women suffer more than men. That is the result of having been under the regulations of long-lasting past social values and norms. The facts above show that social standards giving rise to differential treatment between men and women are still deeply rooted in the hearts of Chinese people and have formed an important part of their self-concept.

This is the crux of the matter. The female is subordinate to the male not only in terms of objective realities like relying on men for their livelihood and being offered lower-level job positions than men. What is more, the subjective thinking of women as well as their construction of the self also follow the long-established model of "men decide, women follow."

This phenomenon exists in women regardless of whether they are local or outsider and regardless of their education level. The stunned reaction of Rongzong Sao and her denial of sexual activities in the village, the way Wang Guizheng conspicuously wanted to demonstrate her clear-cut, strictly-business attitude—both reflect these women's desire to evade the actual situations they were in. What is more important is that almost all of the women I spoke to blamed the female workers from other provinces for local men's extramarital affairs. Even Li Mei, an outside female worker with a relatively high educational level, also consciously or unconsciously blamed women for their indecent behavior in explaining what was happening between men and women and men's sexual harassment in the factory.

This confirms what we mentioned earlier in this chapter. For a long time the regulation of sexual behavior, whether proper or improper, is targeted at women. Men have a large measure of freedom with regard to sex. Whatever restraints or constraints imposed on them are not based on the emotional responsibilities they should have toward women. Rather, they are formed under the influence of traditional values. This is reflected by how men of the older generation like Uncle Ming and how men of the younger generation like Zhichao in Baixiu Village viewed the issue of "buying a mistress." It seems that both men viewed it proper and fair to buy a mistress if a man could give his wife enough money to sustain the family.

If we believe that the characteristics of a generation are clearly reflected in the self-concepts of the members of that generation, then we can see that this holds true for the Chinese throughout the past several thousand years. Very often these self-concepts play an important part in creating one's existential predicament. The problems faced by the local women of Baixiu Village surrounding marriage issues have demonstrated this point well.

Of course, as we mentioned earlier in this chapter, we understand that the women in Baixiu Village, especially those of a younger age, have become increasingly aware of their autonomy during the past ten to twenty years. However, this growing self-awareness is confronted by the traditional "men superior, women inferior" values. This confrontation increases the pain of the women and has generated a kind of suffering not experienced before. The women in Baixiu Village have become clearly aware of their predicament in marriage and related issues; at the same time, they lack the sense of having a designated female role in familial relationships. On the one hand, they can no longer take refuge in being "resigned-to-one's fate" as they did before; on the other hand, they cannot change the objective social conditions and the subjective opinions traditional women held toward sex. As a result, they suffer from a deeply felt melancholy as regards relations between the sexes. To a certain extent, this is a fair portrayal of how Chinese women in general feel toward relations between the sexes since the initiation of the reform and opening-up policy.

13

The Paradox of Development in Baixiu Village

In recent years, there seems to have emerged a widespread common belief, found in daily conversations among ordinary people as well as in books and research monographs, among writers both in China and overseas, that China in the last ten to twenty years has been changing too much too fast. Fueled by the market economy, a person has more jobs to choose from, and this has accelerated social mobility. This phenomenon is especially apparent in villages, and all the more so in the Pearl River Delta. Piece after piece of farmland was leveled to become factory sites. People whose family had engaged in farming for generations have now become factory workers and supervisors, even factory owners and managers. Others have switched to jobs such as transportation and owning small shops; still others became involved in illegal enterprises. The sudden expansion of the number of factories created a huge demand for workers. In many cases, a village that had a population of about 2,000 would suddenly swell to 30,000 or 40,000 people, most of them women workers from other provinces. This kind of development was very common in the Pearl River Delta. The site for our fieldwork, Baixiu Village, is typical.

On the other hand, economic development has spawned many social problems. Corruption and financial scams occurred in different guises; sex parlors proliferated. The extreme instrumental idea of "money first" pervades society. Part III of this book shows that almost all the male residents have extramarital sex. Having mistresses and visiting prostitutes is quite a common occurrence. The sexual activities of the primary school principal, the supervisor, and the teacher are clear reflections of the prevalence of sex.

The relentless pursuit of wealth, the overindulgence in sex, and the deterioration of law and order give people a strong sense of the dissolution of social order. The villagers have more wealth now; individual freedom has been expanded. These positive developments combine with the dissolution of social order to form a distorted society. In fact, this is not peculiar to Baixiu Village. It may indeed be the microcosm of China. What are we to make of this? In particular, how are we to make sense of these phenomena in the context of Chinese modernization? This is a complex question. China's change is extremely rapid, and we are in the vortex of the change. It is not easy to discern the direction of China's development.

Throughout the stories in the book's three parts, we have tried to base our analysis on the actual context of each story, to explore the changes in the village by looking at real happenings. In this chapter, we attempt to probe the development of Baixiu Village from a macro perspective, and shed light on some of the questions encountered in China's modernization process since the reform and opening-up policy. But first, we have to emphasize this: China's development is so rapid and varied that it is very difficult, indeed impossible, to describe it in full. What one chooses to describe, out of the myriad phenomena and the directions of development reflected by them, often depends on the researchers' analytical perspective, the point of entry to the question, and the theoretical framework. We are aware of the limitations of empirical social research itself. We are also aware of the limitations and bias of any theoretical framework in the gathering and analysis of fieldwork data. Therefore, the objective of our research is not to find out the "truth" of the whole picture of China's modern social changes. It is rather to portray and analyze some of the advances and setbacks in the process of China's modernization. This we attempt to do from an angle that is both meaningful and familiar to us. Using Baixiu Village as our base, we employ the method of participant observation "to see the world in a grain of sand."

The Market Economy and the Dissolution of Social Order

Many people are of the opinion that China today shows signs of a clear dissolution of social order and the bankruptcy of traditional values and beliefs.[1] Others think that the present time is the best in Chinese history in half a century.[2] Still others characterize China as experiencing an era in which "there is plenty of vitality in people and much chaos in social order."[3]

Under the rule of the Communist Party after 1949, China underwent tumultuous changes—in economic practices, political system, values, beliefs, and ideologies. Comparing those changes to the ones resulting from the reform and opening-up policy after 1979, one can see that the latter in some ways have a more comprehensive as well as deeper impact on the lives of the people. They can be construed to a certain extent to be changes from a traditional society to a modern society. Such a transformation has been seen in many other societies in the world in the past century. It has been the main subject of discussion of Western social science. The historical developments of China and the West are of course embedded in their respective social and cultural contexts and different from each other. But from a broader perspective, one of the foci of the great changes in China now is the weakening of the influence of traditional values on people and the rise of individual autonomy. Set against our theoretical framework, this means the waning of the influence of familism and the waxing of the sense of individual autonomy triggered by a market economy.

To some extent, the revolution in the thirty years between 1949 and 1978 did not shake the deep cultural structure and way of life of Chinese people. On the surface, familial culture was suppressed. The farmer's way of life went from being a matter of private practice to one of public management. However, after the implementation of the reform and opening-up policy after 1979, the market economy brought in its wake certain values and notions that challenged traditional values and beliefs and affected the choices one had to make in daily living and in matters involving one's values. The development in Baixiu Village exemplifies these changes vividly. We see someone like Uncle Qiu, who believes firmly in traditional concepts of marriage and who enthusiastically revives lineage activities. Even Zhichao, of a somewhat younger generation, is clearly influenced by traditional values in his views about family and his attitude toward his wife. In the persons of Zhichao and the younger Qiming we see evidence of the change in values and beliefs in the transition from tradition to modernity. This is even more evident in the person of Wang Guizhen, who demonstrates how a young woman develops her autonomy through her own efforts in the emerging economic environment of Baixiu Village.

Such changes as documented are very different from those occurring before 1979, including those in the thirty years between 1949 and 1978. It is not our intention to exaggerate the effect of the economy on human behavior. Yet judging by what one can see from the case of Baixiu Village, it is difficult not to ascribe the main cause of the changes to the market economy promoted by the reform and opening-up policy.

Put another way, because of the improvement in economic life and the more relaxed political environment in the last ten to twenty years, the villagers of Baixiu Village have learned to use the resources they acquire to plan their own lives. Certain concepts related to market economy, such as "fair exchange," "private property," and "investment," came to be cited in the calculation of personal interests. We can say that individual autonomy gradually developed in actual daily life in answer to the primary self-serving premise of how to satisfy one's needs and wants.

Conversely, individual autonomy induced by market economy is in conflict with familism on different planes. The various conflicts in daily affairs—between Qiming and his wife on the one hand and his mother on the other, and the quarrels between Dayan Sao and her mother-in-law—can be said to reflect the contradiction between old and new values and ideas. Such conflicts seem unavoidable on the road from tradition to modernity. However, in view of the present state of development in China, many people see such change in values as the source of chaos.

From a different angle, the quality of life of Chinese people after the reform and opening-up policy has been greatly improved. This is a new milestone in the history of China's development, especially for villages. Take two examples: the villages along the Pearl River Delta and those of the Yangtze River Delta. Compared to others, these two regions have fertile soil. Nonetheless, before 1980, most of the villagers living in these two regions had to struggle to subsist.[4] Baixiu Village, our fieldwork site, is near the Pearl River. Other than farming, people there can catch fish and other seafood as a way of making a living. The standard of living is above average in the Pearl River Delta. The villagers there told me, however, that not until the 1980s did they have enough rice in their three daily meals.

We can say that after 1979, people in China not only improved their standard of living tremendously, they also made a big leap in a very short time from a life of subsistence and struggle to one in which they could pay attention to quality of life. Before 1979, only a few people had a television set and an electrical rice cooker, but now almost every family has such electrical appliances. Richer people can even build their own houses. As far as we know, some richer cadres have their houses rather expensively decorated and furnished. Bathroom fixtures could cost tens of thousands of yuan. This affluence is obviously the positive effect of modernization driven by the market economy of the past decade or so. That may be why some people always stress the importance of satisfying material needs.[5] The same reasoning might be behind the frequent remarks made by China's leaders—as a retort to criticisms about the lack of progress in China's human rights conditions—that feeding 1.2 billion people is itself an achievement in human rights.

Our research focuses on the social aspect of these rapid changes. Economically, life has improved, but in other respects such as values and human relationships, there seems to be a regression. In Baixiu Village, it appears that only money and sex matter. Even at a parents' meeting in a school, the principal and teachers request that parents use monetary incentives to promote students' academic proficiency. Of course, no one can deny, as pointed out earlier, the material improvement in people's lives as a result of economic development. Looking at the history of the past half-century, the current prosperity is something to be greatly valued. But is China's development something as sanguine as imagined?

The individual autonomy fostered by a market economy is at first something very positive. Market economy presupposes the recognition of private property. It encourages the pursuit of profit and, through the free trading of goods, promotes the notion of fairness. Individual autonomy can be realized through economic activities. Modern Chinese are tending toward constructing an autonomous self-image.

Of course, we are aware that China's present state of development still lags behind that of the so-called post-capitalist countries, whether Western societies or the more advanced countries in Southeast Asia. However, from such countries it is clear that in development that is guided by instrumental rationality, the pursuit of wealth and a life-goal centered on the self are fundamental aspects of a market economy.

Since the middle of the twentieth century, modernity or modern society has been under attack in Western academic circles. Different schools critique various aspects and lifestyles of modern society. Chief among the targets are the unrestrained expansion of modern scientific technology, the destruction of the environment, human relationships based on material considerations, and the sense of "alienation" experienced in a bureaucracy-dominated social organization.[6] Some scholars say that these flaws of the post-capitalist society will not necessarily appear in Chinese society. That is possible. The problem is that the current dissolution of social order in China is in some ways worse than the flaws listed.

We want to emphasize again that we do not intend to fault the positive value and significance of the development of the Chinese market now by referring to the shortcomings of Western capitalist society, especially since China is a country that has long been burdened by poverty and in which autonomy has long been suppressed. We do want to point out that while China's development is only at an early stage compared to that of the West, many of the problems that surface in Western countries have also appeared in China. In particular, the prevailing general social practice in China seems to be even worse.

Take Baixiu Village as an example. What makes one think of it as degenerate and chaotic are the prospering sex trade and the extremely distorted relationship between the sexes. As described earlier in Part III, we see the astounding degree of sexual indulgence and preoccupation on the part of the male villagers. Almost every adult male villager is involved in extramarital sex or an extramarital love affair. This is quite common in the Pearl River Delta and common too in almost all the economically developed areas in China. There is only a difference in degree, not in substance, among them.

There are maybe two main reasons for the rapid spread of the sex trade in Baixiu Village. One is a concrete reason. The market economy has increased people's wealth, and there has been a massive influx of female workers from other provinces. The other reason has to do with a deeper structure of the cultural fabric. Familial culture has weakened as a principle governing human behavior. The influence of traditional morality has also weakened. In chapter 1, we pointed out that the moral system of Confucianism is less adaptable to rapid social change. It provides few tools for coping with the temptations presented in modern society. The individual shaped by familial culture lives according to a complicated relational network. In this network of relationships, patriarchal culture and the relationship between the sexes characterized by "men decide, women follow" are two of the main features.

If we further analyze the characteristics of such cultural artifacts, we can see more clearly how Chinese self-concept is molded by the Confucian ethical system and familial culture. When we compare this with the self-concept of the modern Western person, we find that modern Western values are to a large extent based on the individual's subjective sense of values and personal interests. Unlike traditional Chinese, Westerners do not rely on the value system of familial culture. Thus, the Chinese find it hard to build up a "subjective values" system based on the individual and outside the moral imperative of the collective. As traditional values disintegrate, a new underpinning for individual values and beliefs has not developed to take their place, leaving a vacuum. The West in modern times has also experienced crises caused by the disintegration of religious systems and the loss of faith, but the West has a stronger sense of "individual autonomy" or "self-concept," and a firm and long tradition of "expressive culture"—witness the vibrant and continual development of the arts. Therefore, although the "dissolution of social order" has occurred, individuals have had more means of expression. By contrast, in China, self-determination or individual autonomy was much weaker, so when the "dissolution of social order" occurred, individuals were at a loss as to what to do.[7] On top of that, when wealth and sex beckoned, the situation got out of control. China in the 1990s faced this situation, and what is happening in Baixiu Village may well mirror what is happening in modern China.

So we ask, What is the direction of the future development of Baixiu Village? Staying within the confines of our research, we are concerned not with the economic development but with social aspects such as marriage and the relationship between the sexes. In this, Baixiu Village presents a complicated picture.

All in all, familial culture has had a weaker influence on daily affairs than before. However, the sex trade has again strengthened and fostered some negative aspects of familial culture, for example, the oppression of people, the oppression of women. It has strengthened the old relationship between the sexes in which men decide and women follow. And the relationship has become even more one-sided. The present relationship between men and women is based on the control and enjoyment of the female body. It can be said that women have become extremely commodified. Many Western feminists have pointed out that the male–female relationship is fundamental to all other relationships. Judging by the relationship between the sexes in Baixiu Village, the oppression of women can be extended additionally to the oppression of the subordinate by the superior in factories and enterprises, and the oppression of the poor by the rich.

Viewed from our theoretical framework, this is the paradox of the "modernization" of Baixiu Village. On the one hand, individual autonomy has greatly increased in matters concerning marriage and daily affairs. On the other hand, the sex-trade phenomenon brings about the oppression of human beings and the deterioration of human relationships. From a wider perspective, this perhaps shows the direction of China's modernization: what it can achieve, what confounds it, and what the stumbling blocks are.

The Tension Between the Development of Autonomy and Familism

Valuing individual autonomy and valuing the interdependence between human relationships are two ways of life. It is not easy to say which is better. These two ways of life are not abstract concepts. They are not lifestyles clearly definable at the theoretical level. They have to be put in the context of particular historical and social developments in order to be comprehended. In this research, we simultaneously use two value systems to examine the development of Baixiu Village and to assess its conditions from its existing social context. We will now inspect the conflicts and influence of these two different value systems and ways of life in Baixiu Village or, perhaps, in the whole of China.

"Modernization" encompasses economic, political, and social spheres. As far as the economy is concerned, most successful modernized regions practice a market economy. One of the key elements of a market economy is the central position of private property. Any economic transaction is premised on the individual's consumption and use of resources. This economic model emphasizes the autonomy of the individual in economic operations. In politics, the situation is even clearer. The purpose of democracy is to protect the right of individuals to choose in different spheres of life. Therefore, market economy and democracy complement each other and facilitate the growth of individual autonomy. If China's modernization is advancing in a similar direction, then using "autonomy" as a theoretical construct to understand modern China is justified.

This raises a problem. If we accept "autonomy" as a theoretical framework in our research, then to some extent we also accept the value judgment that "the development of autonomy is something worth promoting." Examining China's indigenous cultural context, one can see that traditional Chinese values suppress the growth of individual autonomy. This is the crux of the matter. One of the basic tenets of modern Western society is the respect accorded an individual's rights. This tenet goes hand in hand with the development of its material culture. Since China's traditional culture hampers the development of individual autonomy, and China has been weak in the past hundred years, many recent and contemporary Chinese scholars have used the values and beliefs of the West as well as its social progress to attack traditional Chinese culture. They reject China's values and beliefs, human relationships, and ways of doing things.

We think that such an attitude is to some extent imposing the straitjacket of Western ideas onto Chinese conditions. We employ the notion of "the development of autonomy" to understand Baixiu Village, but are against using this notion to criticize everything that happens in Baixiu Village. Even where the criticism is relevant, we carefully state the various kinds of restrictions and qualifications and indicate our own limitations. There are good reasons for doing so. First, we are facing complex and profound questions, and we still lack a clear understanding of many events. Under internal and external limitations, we often can only employ this framework to sharpen some specific questions and use it for some preliminary observations. Second, it seems that few people would oppose the development of individual autonomy, but it also appears that many people do not regard individual autonomy as the highest ideal. In fact, individual autonomy does not mean simply advocating that individuals have the right to make free choices in daily affairs. Behind the concept lie some profound values that influence our way of life in areas that might appear unrelated.

Regarding autonomy as the supreme right of humans connotes a moral inclination, that is, that an individual's feelings are the ultimate criteria for discerning "virtue" and "evil," "right" and "wrong." From this point of view, moral standards are to a large extent based on a person's "feelings." People do not only express their feelings, they attain the ideal of "selfhood" through them. In so doing, they also actualize a moral self.[8] Viewing human relationships, social organization, and political institutions from this moral perspective, we can see the formation of a social and cultural context in which individuals take center stage and individual feelings are externalized ad infinitum.

If we adopt the value system and corresponding human relations network discussed above and use it to measure the Chinese views of the individual, family relationships, and social structure, then everything about China seems to be full of faults in that it restricts the nature of human beings. We do not deny that cultivating individual autonomy spurs the development of modern civilization; we admit that to a large extent traditional Chinese culture suppresses individual autonomy. It does not follow, however, that we identify with all the human relationships engendered by that belief, nor do we accept that individual feelings are the final arbiter of morality.[9]

Does advocating the development of individual autonomy inevitably run into conflict with traditional Chinese culture and familism? The answer to this question depends on different historical developmental stages. The answer will be different accordingly. Looking at the period of the 1990s, we can say that the answer is "no." Traditional Chinese culture and the ideas of modern society are not completely incompatible; nor are conflicts between them inevitable. However, many scholars who look at the transition from traditional to modern society are more inclined to answer "yes." Scholars of the transitional period of time think that traditional China and the modern world are incompatible. We shall examine this question in a broader perspective.

Traditional Chinese culture has unique characteristics, but it also shares some features with other cultural traditions. We can explain this by using Emile Durkheim's distinction between "mechanic solidarity" and "organic solidarity."[10] The former refers to the structure of traditional society. Its distinguishing feature is a lack of division of labor, but it has a common set of values and social standards as well as trust in an authoritative leader. The latter is modern society with its characteristics. The division of labor is present to a fine degree. There is a lack of convergent values and beliefs, and actions are taken on the basis of individual choices. Obviously, these characteristics are applicable to traditional Chinese society and its modernization process, with all the difficulties inherent in the move from one to the other. In other words, the process of Chinese modernization can be seen as a move from "mechanic solidarity" to "organic solidarity." Therefore, the authority of the family is in conflict with autonomy and its influence on people is on the decline. The Chinese people become materialistic. Such attitudes appeared in Baixiu Village, and are a universal characteristic of the transition from a traditional society to a modern one. When we view things from this angle, we cannot put all the blame for mistakes and problems on traditional Chinese culture, and we will not see such aberrations as totally negative social phenomena unique to Baixiu Village.

On the other hand, we admit that Chinese culture has unique features that many people blame for the dilemma China faces on the road to modernization. They think that traditional Chinese culture or familism cannot coexist with the value system and ideology of modern society. As stated earlier, we can discuss this question in relation to two historical stages. From the Opium Wars in the mid-nineteenth century until 1978, Chinese traditional culture was in violent conflict with ideas associated with modern society. Chinese reaction to Western ideas was extreme, Chinese modernization was beset with failures. But if one examines the Chinese diaspora in Hong Kong, Taiwan, and Singapore in the past decades, along with China in the past two decades, then it is obvious that Chinese and Western values and ideas did not clash violently. Chinese people in these areas have been able to make use of Western ideas to develop self-awareness and further political and economic progress.[11] Inevitably, though, Chinese and Western values come into conflict in various ways and such conflicts influence people's lives. But the era of extreme reactions is over. In Baixiu Village, we have seen how villagers are affected by the market economy and how it improved their lives. We have also seen how people like Wang Guizhen and female workers from other provinces like Li Mei developed a strong self-concept when they came under the influence of values engendered by a market economy. At the same time, we see in them the influence of Chinese familial culture.

In the stories narrated above, we have stressed that familial culture restricts human behavior. At the same time, we see how familial culture cares about the moral side of human relationships. The parents in Baixiu Village normally provide money for their children's marriage and build a house for them. Friends help each other in monetary and other matters. These are signs that familial culture is concerned with mutual help. To a certain extent, the residents of Baixiu Village maintain such human relationships, while gradually reducing the reliance on authority. This shows the influence of individual autonomy. At the same time, it demonstrates that the emergence of individual autonomy can go hand in hand with the emphasis placed on human relationship as emphasized by familism.

To return to the theoretical framework about the characteristics of familism described in the first chapter, we can see the positive side of the human relationships generated by the "father–son dyad." The relationships exhibit the characteristic of continuity, in contrast to the instrumentality of relationships based on personal feelings. A long-term, continuous human relationship puts more emphasis on mutual understanding. It is true that relationships evolving out of the "father–son dyad" are based on an authoritarian model in which there is a top-down chain of command. But from what can be seen from the current state of development in China, the relationship described in the proverb "the seniors are exalted; the juniors are despised" is getting weaker by the day.[12]

Maybe we can see reasons for being optimistic about the future development of Baixiu Village, but there are also reasons to be pessimistic. People pursue the end of wealth without caring much about the means. The sex trade is pervasive. Women are extremely commodified. However, there is strong conflict between tradition and modernity, between Western and Chinese ideas. The villagers' standard of living has been greatly improved. To a certain extent, people have maintained relationships as espoused by the traditional value system. If we examine the positive impact of that value system on traditional Chinese culture, we can see that, if such features get properly developed, they will imbue people with a sense of "ultimate concern"—to a greater extent than will a morality built on individuals' feelings. It may to some extent lessen the sense of alienation among people living in modern societies and help to restore meaning to their existence. Of course, we need to be careful not to overstate the status of Chinese culture in the modern world. We merely point out that the Western value system, which centers on individuals' feelings, is seriously flawed in regard to the levels of human existence and morality. The ethical system of human relations as inherited by the Chinese people may offer some guidance to today's world as it grapples with disjunctures between the traditional and modern social environments.

14

Reflections on Methodology

This chapter will reflect on the methodology that we use in this study. We mainly used the method of participant observation to collect data, the perspective of critical hermeneutics, and the method of thick description to present and analyze the data. This methodology has been in existence for a long time and is widely used in social science research in the West. In Chinese social science circles, it is used primarily in the field of anthropology. Therefore, in this chapter, we go into greater detail to explain the research methodology. Our aim is to combine the methods of thick description and critical hermeneutics and then to sketch some of the problems related to indigenous social research.

The Positivistic Conception of Science and Empirical Social Research

For many people, one of the main reasons why the social sciences merit the appellation "science" is that the results of social research can be verified as to their truth or correctness by empirical data gathered from the real world. In other words, the features of "science" in the social sciences come from the possibility of conducting empirical social research. Such empirical social research generally regards the positivistic conception of natural science research as a paradigm for carrying out investigations. That is, many social scientists, especially those specializing in empirical investigations, as well as authors of textbooks on social research, try to use the natural science model to teach and design social research.[1] They use concepts developed in the natural sciences, such as "verification procedures," "generalizations," and "replicability," to evaluate the findings of social inquiries. But because of the use of these concepts, there are problems in the development of the social sciences, especially sociology. What kind of results can be obtained from social research by following or imitating natural science models? Sociology is often regarded as a dull and superficial subject, and the results of social research are scorned for being only glorified "common sense." This is because many sociologists attempt to follow the research procedures and goals of the natural sciences.[2] Sociology or social research is faulted because it accepts the positivist conception of natural science as its model of social investigation.

The chief reason why natural science has been regarded as the paradigm for knowledge is due to its rigorous verification procedure. This rigor is attested to in the accuracy of its predictions. For a positivist, there are three assumptions underlying natural scientific inquiry:[3]

1. The truth or falsity of a theory is dependent on how well it corresponds to the world it describes. This is called the "correspondence theory of truth."
2. There exists an independent objective world. This is "simple realism."
3. Human perception has "objectivity" or "intersubjectivity."

These three assumptions are linked to another assumption: the existence of "basic fact."[4]

Basic fact is the fundamental unit of complex phenomena. The researcher need not interpret it but can understand it purely through perception. The basic unit of natural science theory is a "basic proposition." A basic proposition describes or denotes a corresponding basic fact. We can determine whether a basic proposition corresponds to a basic fact through sense perception. Because human sense perception has objectivity or intersubjectivity, there is an objective basis for the verification of basic propositions. The verification of basic propositions can be extended to the theoretical level, so the truth and falsity of a natural science theory can be objectively verified.

The above assumptions raise many questions. First, are the three basic assumptions of natural science valid? Second, do basic facts exist? Third, even if the answers to the above two questions are positive, can they be applied to social research? In discussions in the field of the philosophy of science today, very few people accept a positivist view of science, and much less a positivist view of social science.[5] However, the positivist model of research has been so entrenched in empirical social research, especially in Chinese academic circles, that it is necessary to discuss it together with social research.

The question of the nature of the natural sciences is still beset with controversy, but we can use the "pragmatic criterion of predictive success" to justify the objectivity of natural science research.[6] In other words, no matter what we take to be the nature of the natural sciences, to a large extent its accuracy of prediction can demonstrate and support the objectivity of natural science.

On the other hand, although few people deny the high predictive accuracy of the natural sciences, it does not mean that basic facts exist. From the perspective of post-empiricism, the truth or falsity of a theory or postulate does not depend solely on its correspondence to the outside world. Rather, to a large extent it is decided by the "consensus" of the involved academic community. This has been a focal point of discussions of the philosophy of science in the past half-century.[7] As this issue involves complicated questions, we shall not discuss it further here. But since it is related to how we correctly understand the nature of social research, we shall use the following example to explain the truth or falsity of a theory and its relationship with "consensus."

Suppose there is a proposition or theory X1, and X is the empirical world that this proposition or theory refers to. The truth of X1 is not dependent entirely on X, which it denotes, but also to a certain extent on how we interpret and understand X. We may come to a consensus about the content of X1 and consider it to be true, but once new evidence emerges, we will have a new interpretation of the empirical world that X1 refers to. Such a new interpretation may change our theory X1 into X2, but our empirical world X has not changed. That is to say, the world X to which X1 or X2 refers has not changed. Only our interpretation of X has changed. This means that the X that is the object of our experience is still the same; what has changed is our interpretation of it. In the history of science, we make progress in our understanding of the natural phenomena that we experience by making new interpretations about them. It is not the case that all progress made in scientific knowledge is marked by a change in the experienced world. An obvious example in the natural sciences is that our sensual experience of the celestial bodies has not changed. What has changed is our interpretation of such experience. New interpretations lead to a modification of old theories or the emergence of new ones.[8]

This shows that even in natural science research, the so-called basic fact that is perceived by the senses can only be understood with the addition of the interpretation of the researcher. Thus, basic facts do not exist at all. If this is the case, then how is the objectivity of the natural sciences established? This is a complex question. All that we would like to point out here is that no matter what view we come to take about the nature of the natural sciences, its high predictive accuracy is sufficient grounds for supporting and verifying its objectivity.

If we turn to social research, we will find that quite a number of research projects start off with a mistaken positivist view of science: to try to find a related basic fact and to hope to achieve a "rigorous" research procedure or result. Employing a questionnaire method in quantitative social research is an example of this research model. It imitates the natural science model in every way, from the stage of postulating research assumptions, through constructing theories, to gathering data and analyzing them. By doing this, it is hoped that a "rigorous" verification procedure will result. It attempts to use the interview with the subjects, that is, the speech and words of the interviewees, as the basic data of the research.

In reality, this process exhibits two kinds of misunderstandings about the social sciences on the part of the positivists. First, as we said earlier, there is no basic fact. Second, there is a difference between social phenomena and natural phenomena. Although both involve the researcher's interpretation to some extent, the latter do not have a tight or close relationship with social standards and values. Social phenomena are made up of elements such as linguistic symbols, customs, values, and rules. It is more difficult to establish basic facts in social research. This is also the main reason why it is difficult to achieve predictive success with social research.

In light of the brief analysis above, we can further examine the so-called rigorous research or investigation procedure of the positivists. Generally, the ideal research model for positivists is quantitative social research, especially the questionnaire survey method. They will use the criteria in the questionnaire survey method to interpret and assess the validity and reliability of other research methods. They will also criticize nonpositivist social research for lacking a scientific basis. What exactly are the "rigorous" procedures and criteria of the positivists?

There is a rather popular misconception in social research. Many people think that the quantitative social research method is an exclusive method for positivist social research. They think that nonpositivist social research will use only qualitative methods. This confuses positivism with an empirical method. The former is a school of social research; the latter refers to a research method based on empirical data. In fact, all social research bases its research on empirical data and uses the empirical method to collect data. Both qualitative and quantitative research methods, including questionnaire surveys and participant observations, are methods for collecting and analyzing data and are available to any school of social science. Thus, quantitative social research or questionnaire survey methods are not the preserve of the positivists and cannot be used to prove the rigor of positivist social research.

It has been pointed out that the rigorous research model favored by the positivists is based on their objective verification procedures. According to the above analysis, the so-called rigorous verification procedure is actually a set of assumptions about social phenomena. The assumptions in turn are built on certain understandings about the nature of the social sciences. Therefore, the three basic assumptions of the positivists can be construed as constituting the "rigorous" objective verification procedure. But in this sense, any assumptions concerning the nature of social research become a "rigorous" verification procedure.

We may now come to a tentative conclusion. The positivist view of social research is based on a mistaken view of the natural sciences. It therefore leads to a misunderstanding about the nature of social research and the process and techniques of social investigation. This mistaken understanding is extended to all social research methodology.

Explanation of Social Phenomena and Qualitative Social Research

One of the main aims of social research is to achieve social explanations. This involves two levels of explanation: to try to find out the causal relation in a social phenomenon and to make a thick description of that phenomenon. Both positivists and nonpositivists agree on this aim. Their difference lies in which explanation is basic and what constitutes a thick description.

Scholars engaged in empirical social research generally know that a good deal of practical social investigation involves in-depth descriptions of the situation under study and not a search for a causal relationship. But positivists think that a description or explanation of any social situation faces the problem of verification. Even if the explanation involves factors so complex that they cannot be verified at the level of experience, the researcher will still attempt to design the research process in a way compatible with the empirical verification procedure. In other words, for a positivist, ideally and at the level of verification, the thick description of a social phenomenon will have to be restated ultimately in the form of an explanation of a causal relation; otherwise it will not be objectively verified and the study cannot be said to be scientific.[9]

This view of the relationship between a causal explanation and thick description leads many researchers to misunderstand the meaning of thick description and to become confused about the notion of social research. One example is in social research that makes use of in-depth interviews; the form of the interview, its conceptual design, data compilation, and analysis follow the format of a questionnaire survey and neglect the characteristics of social research. The researcher consciously or unconsciously follows a mistaken idea about rigorous social research and sacrifices the distinguishing features of the in-depth interview. This error is the result of a misconception about thick description.

Thick description differs from a causal explanation in one important respect: the latter emphasizes the cause-and-effect relationship of a social phenomenon and tries to achieve an accurate prediction of the situation, whereas thick description attempts to present the many characteristics of a social phenomenon embedded in its relevant social and cultural context. However, social research has been coupled with the term "science," and for many people, science involves objective verification and predictive success. Therefore, many people unconsciously explain social phenomena from a cause–effect perspective. Then the question arises: If thick description does not aim at predictive accuracy and is not objectively verifiable, then what kind of message can be conveyed in the social explanation it offers?

This is not a simple question. The general public sees scientific explanations in the natural sciences as a matter of tracing cause and effect, as we pointed out earlier. In fact, in the natural sciences, one seldom links the "scientific" explanation of natural phenomena to related social and cultural contexts so as to achieve a thick description of the phenomena concerned. There is a difference between the language used in natural science discourse and that used in social science discourse, and in the relationship between this kind of language and everyday language.

The language used in natural science, a specialist variety, differs in nature from the language used in everyday life. The language of social research, on the other hand, is our daily language. A "scientific explanation" of social phenomena and an explanation of everyday events both operate at the same level. There is a difference in degree but not in kind. From this point of view, we can further explicate the concept of a social explanation, especially that of thick description, in terms of our daily understanding of human and social relationships. This is precisely the viewpoint of "phenomenology" or "hermeneutics" in the nonpositivist school of social inquiry.[10]

In our daily life, we often think that we have come to an understanding of an event without saying that we can pinpoint any cause–effect relationship. What we mean by "understanding" is that we have come to see the event in the social, cultural, or value context in which it is found, or that we have a deeper sympathetic understanding of the event or persons involved. Some people might not accept this kind of understanding as the goal of social inquiry because it is at a "noncognitive level" of understanding. But their objection involves a definition of "knowledge" and the relationship between "explanation" in social inquiry and everyday "explanation." If we accept the thesis that there is a commonality between the two kinds of explanation, then we will find that, to a certain extent, the aim of a great deal of social research, such as case studies and participant observations, is to achieve a thick description. However, due to a mistaken notion, many people use a causal model and concepts to design and direct the conduct of social research.

The above discussion involves two important issues: the nature of social reality and value-involved social research. In fact, thick description has become an important goal of social research mainly because social fact has one feature very different from happenings in the natural world—social fact can be understood as an "institutional fact." In other words, social phenomena are unlike natural phenomena; they do not have an objective existence outside of humanity. Any social phenomenon stands in an inner inextricable relationship with the related value system or social rules. Social rules are constructed and transmitted mainly through language and language carries a communicative or intersubjective meaning. This meaning not only makes communication possible, it also constitutes the social life of human beings. When we use language to describe the social world, we are also using it to construct or reconstruct the social world. That is to say, language is not just purely an instrument used to describe an independent and objective social world; it is an important element in the construction of the social world.[11]

According to the above analysis, the language an investigator uses to understand social phenomena and to collect and analyze data is not purely an instrument of social research. It influences and even constitutes the phenomena he or she sets out to investigate. The use of language, of course, has certain universal properties, otherwise human communication would be impossible. At the same time, we have to acknowledge that different people using the same language or vocabulary will use it according to their own meaning and usage. The language of the researcher will consciously or unconsciously intrude into one's investigation. Insofar as the research through his/her language intrudes, the values of the researcher will intrude as well. That is one reason why it is difficult to establish a "causal explanation" in social research.

How does the involvement of values affect practical social research? How does it affect thick description? The involvement of values in social research is a controversial issue.[12] There are two levels of value involvement, which will be described below. From this description, some practical questions in social research will emerge.

The first kind of value involvement is based on the principle of "value relevance."[13] In social research, the social scientist chooses the object of study according to some criteria or framework and uses these to assess the impor-

tance and significance of the research question. These criteria and framework become the "value relevance" of the study. For sociologists like Max
Weber, the "value relevance" is not the same as the researcher's own "value
judgment." The value system implicit in the "value relevance" is different
from the "value judgment" of the researcher; it is made up of the values
generally accepted by the community to which the researcher belongs, and,
as such, it has an objective significance.

Weber further points out that the involvement of value in the "value relevance" will not affect the objectivity of the results. After the research question has been selected, then that becomes an objective standard, guiding the
collection of data and their analysis. The truth and validity of the research
results are determined by the objective outside world and the methods of
social science.[14] To quite a large extent, Weber's viewpoint has been accepted
by positivist researchers and it has found, perhaps implicitly, its way into
general sociology textbooks.

If we take "institutional fact" as our vantage point, we will find that value
involvement is not confined just to the selection of a research question. In
fact, Weber himself clearly points out that the main purpose of social inquiry
is more to understand the meaning of social behavior than to track down
cause and effect. Researchers have to connect social behavior with its relevant cultural tradition and values to arrive at the meaning of the behavior.[15]

It has been pointed out that the inclusion of values at that level is not in
conflict with the value-neutral verification procedure, since after the research
question has been selected, the research steps that follow can be guided by
the research question and the objective facts of the outside world, thereby
ensuring the objectivity of the research. The fact is, though, that Weber emphasizes that the goal of social research is to arrive at an in-depth understanding of social phenomena. Furthermore, this in-depth understanding
comes from connecting the phenomena with the relevant value and cultural
systems. In this case, how can the verification procedure in social research
be unaffected by values?

Compared to the viewpoint of the "value relevance," the "strong thesis of
value involvement" emphasizes that in social inquiry, researchers will, consciously or unconsciously, include their value judgments in their research.[16]
In contrast to the traditional positivist view, this kind of value involvement
encourages researchers to take stock of the relevant situation and to merge
their value judgments with the theoretical framework of the research. A prominent example is research on the conditions of women in modern times. According to many researchers, if one does not adopt the perspective that
criticizes patriarchal dominance in modern society, then one will not have a
correct understanding of women's condition.

A question naturally arises: If researchers have to construct a theoretical framework containing their own value judgment and they apply that framework to research, how can they ensure its objectivity? Even if we take thick description to be the goal of inquiry or explanation, acknowledging at the same time that thick description cannot be confirmed objectively, how can we know whether it is reliable?

This difficult question perplexes nonpositivist researchers. We cannot fall back on the traditional concepts of "objective perspective" or the "truth" for an explanation. We first have to understand the argument of "the strong thesis of value involvement," then combine this argument with an understanding of an analysis of social phenomena, and finally examine some relevant specific questions in social research.

Compare the following two sentences. "Washing the hands five times a day is noble behavior" and "Tolerance is noble behavior." The literal meanings of these two sentences are clear. But in the absence of any explanation from the speaker, the first sentence is puzzling, for we are hard-pressed to think of any "reasonable" argument to support the assertion that washing the hands five times a day is noble. On the other hand, even if we do not subscribe to the belief that tolerance is noble behavior, we can envisage the speaker putting up some "reasonable" arguments to support the assertion. Or, to put it another way: we not only understand the literal meaning of the second sentence, we also understand the intention and reason behind it.

The above analysis shows that an understanding of a sentence or of a human act is closely linked to the argument behind it. To understand a human act, to some extent we must be able to figure out a "reasonable" argument to support it. We can conclude that one who understands an action accepts that the argument behind it is a "reasonable" argument. A "reasonable" argument need not be sound; it need only be such as to enable one to find reasons supporting that action in the known cultural value system.[17] A person who makes an assessment as to whether an argument is "reasonable" or "unreasonable" is undoubtedly making a value judgment. In this way, we can say that an understanding of social or human phenomena entails a value judgment.[18]

Combining the above with the concept of the "value relevance" explained earlier, we can come to a straightforward conclusion. Someone who seeks to understand social phenomena or human behavior must assess the argument behind them in order to comprehend their meaning. The assessment of the argument behind the phenomena and behavior as well as the understanding of their meaning will have to be carried out against a relevant background, that is, the relevant cultural and social context and value system. This leads us to the following question.

Up to now, our analysis of the "value relevance" has concentrated on the perspective and meaning of the person who seeks to understand. But at the same time, understanding also involves being understood. The subjects of one's study have their own social cultural background and value system. Therefore, the "value relevance" should take into account the two cultural contexts and value systems. Often, the researcher and the subject(s) live in the same social cultural context. However, social cultural contexts can be viewed at the micro or individual level. That is to say, people who live in the same society with the same culture develop individual value systems and meaning contexts.

From the above analysis, the "value relevance" can be the meaning context developed by the seeker of understanding and the understood as individuals. The process of seeking to understand involves the understanding of the two meaning contexts of the two parties concerned. The individual makes use of his or her own meaning context to understand the meaning context of the other side. Therefore, any understanding of a social or human phenomenon is achieved through an adjustment of the two meaning contexts. It follows then that a better way of constructing understanding and compromise is for the one who seeks to understand and the one to be understood to engage in dialogue and mutual interpretation. In other words, we can say that "understanding" is a two-way dialogical process of communication built up by both the seeker and the sought. It is also a dynamic and continuous process.[19]

We may now see how the interpretation of social phenomena is inextricably linked to related value judgments. In social research, the delineation of social phenomena is carried out against the relevant value system and meaning network. A thick description serves to portray all these relevant factors and link them up so as to present the human condition, social problems or oppressions, and inequalities.

Following this line of reasoning, there is no "objective standard" such as those advocated by the positivists to determine the accuracy of a thick description. The accuracy comes from a dialogue among the participants in their social, historical, and cultural contexts. It is here that we see the relevance of our earlier discussion on consensus. The accuracy of a social description is determined by a consensus among the participants, and this consensus can be changed as new interpretations arise.

Hermeneutics and the Indigenization of Social Inquiry

The major difference between the viewpoint of the authors of this book and that of mainstream social research should now be clear. We think that the researcher's value involvement, rather than hindering, helps to understand social phenomena. If one accepts that an understanding of human behavior implies an understanding and assessment of the argument behind the behavior, and that an assessment of the argument must be done by the assessor in the first person, then at least in theory we have shown there to be an intrinsic connection between understanding social phenomena and the value judgment of the person who attempts to understand them. One might ask: Is social research possible without the involvement of value judgment?

In everyday life, very often there is no assessment of relevant arguments. In fact, the interlocutors have presupposed a consensus, accepting each other's arguments underlying the substance of the communication. However, once one person has doubts about some points, for example, by raising a query about the truthfulness or appropriateness of the content of the conversation, then communication becomes discourse. Both parties will have to marshal arguments to carry on the discussion.[20]

This analysis has a special meaning when applied to social research. In our present human condition, social research or social investigation is usually directed at some "problem" phenomena. These can be unresolved points of knowledge, the breakup of the social order, or a personal problem. In other words, to a large extent social research presupposes the existence of "nonalignment" or "conflict" in society and presupposes discussion and criticism of the relevant reasons for such conflicts. To put it another way, whether in communication between people or in social research, there is an implicit aim to reach a consensus or to solve a problem. We might say that behind such human activities lies the lifestyle people hope to achieve, as well as their value orientation.[21] From another viewpoint, this might be understood as the normative basis for social inquiry, as well as the basis for constructing a value-loaded theoretical framework for social research.

It is on such bases that we establish our theoretical framework and conduct our data collection and analysis. In the case of our research, there arises another important topic for discussion, and that is the indigenization of social science. This issue comprises questions at two levels. One is methodology; the other is the more practical question of value judgment. The first is the question of applying Western theories to analyze and understand conditions in China. The second concerns the use of the Western concept of "autonomy" to criticize familism. The following question must be addressed: Will criticizing China through Western eyes lead to "straitjacketing" and unfairness? On the other hand, from a positivist perspective, the two levels of understanding and criticizing are unrelated methodology-wise. But in hermeneutics, understanding social phenomena and assessing their rationality are inseparable. In the context of the stories narrated in this book, the fusion of understanding and critique and of Western concepts and indigenous perspectives can be seen in the relationship and tension between the feelings of the individuals on the one hand and the social structure at the macro level on the other. One of the important points of entry for this research is the interpretation and understanding of selfhood. We are concerned about individual feelings, about their awkward condition of existence, and about the relationship between individual feelings and the corresponding context of the times and value systems. In the following, we shall try to deal with the issue of the indigenization of social research in light of the above discussion.

Few people object to the proposition that one of the most important elements constituting a social phenomenon is a person's behavior and feelings. Hence, a social explanation cannot ignore the micro level of the individual. That is to say, an understanding of social phenomena is inseparable from an understanding of the phenomenon of individuals. And an understanding of the phenomenon of individuals further implies an understanding of the individual concept of the self and the construction of selfhood. Social inquiry, or sociology, deals with social life, so its concern with individuals does not mean individual particularistic feelings, but what is common to individuals within the social group norms or regulations. In this way, what we mean by concept of the self refers to the construction of selfhood or self-identification as generally accepted by the social group. What needs to be pointed out is that we are not neglecting the effect of the macro–social structure in explaining social phenomena. Rather, because of the requirements of the inquiry, we often use the macro–social structure to stand for micro-level individuals, as a starting point for explaining social events.

Whether one should explain social life through the macro–social structure or micro-level individual feelings is an issue that has been debated over and over in social science.[22] In more recent years there has been a tendency for social investigations to combine these two perspectives.[23] From the viewpoint of hermeneutics, the micro level of the individual has priority. One of the reasons for this is that if the understanding of social phenomena implies criticism of the restriction on humans placed by unfair social structures, then the relevant individual feelings or dilemmas can become an argument or standard to assess the rationality of the social structures. This is one of the main reasons why we use the concept of the self as a point of entry for our research.

We can make the following assumption. An understanding of the individual or the self implies an understanding of the moral situation or moral predicament of the individual. When a researcher attempts to understand his or her own self-concept or that of another person, he or she inevitably becomes involved in a relevant moral value orientation.[24] Human beings differ from other animals in that they have the ability to judge values—in particular, they have the ability to judge whether a behavior or thing is virtuous or evil, right or wrong. Very often, a person's distinctive character or self-concept is revealed in choices made in moral dilemmas. We can even say that a person's choices made during the clash of values or in moral dilemmas or predicaments gradually form the character or self-concept of that person as he or she makes life choices. Therefore, to understand individuals means to understand the impinging moral predicament. Yet, as pointed out before, the emphasis in social inquiry is on the moral dilemma or predicament at the level of the social group with the purpose of understanding the generally accepted self-concept of the social group. Each era carries its own characteristics; such characteristics constitute an important part of the self-concept of the people of that era. Such characteristics do not have an objective existence in space and time like ordinary objects. Instead, they are intertwined with some values and rules. We can say further that an understanding of the individual construction of selfhood in a certain era involves as a matter of course the moral value orientation of that era.

It is in this area that we need to be very careful in dealing with the relationship between individual feelings on the personal level and the explanation of social phenomena in a Chinese society in Baixiu Village. Most of the time, we cannot put on Western lenses and attempt to understand the social conditions in Baixiu Village through the individual feelings of the people there. For example, we cannot explain the marriage of people of the older generation, Uncle Qiu for instance, by recording his feelings of love and hate toward his wife. The reason is that their individual feelings toward their wives are not the principal reasons for their marriage. They seldom use personal feelings as a reason for continuing or terminating a marriage.[25]

In other words, we cannot follow Western views and attempt to come to an understanding of the development of Baixiu Village and its social structure by concentrating on the level of individual feelings.[26] This does not mean that such feelings do not make a good point of entry in the research. The crux of the problem is how we proceed from a hermeneutic angle, taking individual self-concepts and their existing dilemmas into account, to combine Western theories and the indigenous cultural context to conduct social research.

Although we use the individual as an important anchor for explaining social phenomena, the relationships between the two planes of the individual and the social are multifaceted. In the social system and moral code of modern Western society, individual feelings are, to a large extent, the bases for individual and social behavior. That is to say, individual feelings not only direct a person's behavior, they also are the raison d'être of one's value or morality. Thus, although Western social systems or social phenomena may mold the behavior of individuals, we can say that they evolve from the feelings of individuals. Individual feelings then become the criterion for assessing the "reasonableness" of social systems, policy, and relevant social phenomena. In other words, this is the view of the individual in Western society, and of the relationship between the individual and society. It is also an important assumption in Western social science. In this sense, to say that one understands society by understanding the individual means to refer to individual feelings. When applying Western social science theories to explain Chinese social conditions, it is easy to adopt the Western viewpoint and use individual feelings as the basis for researching Chinese social situations. But this will be a straitjacket.

We have to be careful. We are not saying that we cannot use Chinese "feelings" as a means to understand social situations. Although Uncle Qiu's marriage is not based on his feelings of love and hate toward his wife, this does not mean that we cannot get a deeper understanding of his marriage through his personal "feelings." We have pointed out above that the relationship between individual feelings and social phenomena is not only that characterized by modern Western development and explanations. We also said that self-concept and its existential dilemma are good entry points for the study of society. Traditional Chinese do not base their construction of selfhood and behavioral rules on personal feelings, but on the concept of the "larger self." That is not to say, though, that the Chinese do not have individual "feelings." On the contrary, we can empathize with the feelings and sorrows of Uncle Qiu, Rongzong Sao, Wang Guizhen, and Fang Ling. We can see how their existential dilemma is influenced by the accompanying circumstances of the social system and norms. These constitute an important part of their self-concept. This shows that their combined self-concept, individual feelings, and existential dilemma differ from those in the West precisely because they do not rely on their personal feelings to either accept or jettison the social norms and phenomena concerned, thus creating their own horrid existential dilemma.

This is an important entry point in the study of these people. Hermeneutics and indigenized research take advantage of this. They take the self-concepts of individuals and use their interpretations and dilemmas to understand and interpret the conditions in Baixiu Village. At the same time, they take the self-concepts as criteria with which to assess the "reasonableness" of the situated social system.

From another point of view, that of indigenized research, we recognize that the relationship between the individual feelings of the Chinese and their social conditions is different from that in the West. It is a different model. To put it briefly, the Chinese people have a different structure to their self-concepts and different existential dilemmas. To see this, it is necessary to start from the traditional cultural context of familism and human relationships. This does not mean that Western values and social science theories do not have a part to play in China research. In the development of China, individual autonomy is rising all the time. More and more emphasis is being placed on individual feelings. There is more awareness too of the suppression of individual feelings and of the relationship between this suppression and the existential dilemma. In other words, to use the concept of the self and the dilemma of individual existence as an entry point, we can come to see how villagers like Uncle Qiu are fettered by traditional views of marriage. Wang Guizhen's problems in finding a mate are constrained by the objective social development and the traditional view of "men superior, women inferior." In this research then, we use the constraint society places on human behavior to understand the conditions in Baixiu Village, and, taking this as a basis, we use individual feelings and existential dilemmas to assess the "reasonableness" of a social system. Thus, Western theories and systems can act as a foil and a yardstick for an understanding and assessment of the Chinese situation.

Part IV

Epilogue: Baixiu Village at the Turn of the Century

15

Baixiu Village at the Beginning of the Twenty-first Century

Our research ended in 1995 and our book was completed in 1998. Since then, members of the research team have been returning to Baixiu Village periodically to conduct further research. With the door between China and Hong Kong now wide open, the tourist trade has thrived. On several occasions, some of our informants in Baixiu Village whom we knew quite well came to visit us in Hong Kong. In conversations during sightseeing and recreation, they told us that Baixiu Village has undergone many major changes in the past few years. The visiting villagers were well dressed and appeared less rustic; the younger women were dressed fashionably. Six years had passed since 1995. It was time to organize the information we gleaned during this period so as not to miss the changes that have occurred to the village and its people at the turn of the millennium.

In January 2002, my colleagues and I returned to Baixiu Village to visit some of the villagers. After settling in the hotel near the village, we took a taxi to Baixiu. The taxi bypassed the entrance to the village and took a newly constructed highway to the industrial area. In the past, the roads in the factory area were made from cement mixed with water and mud. A gust of wind would whip up a whirl of dust even before one could cover one's nose. Now the roads are covered with asphalt, and are wide and straight. From the taxi we could see a small cinema and some restaurants had been added to the old industrial area. The number of factories had increased to over seventy. The park had also been enlarged and reconstructed. The formerly shabby sports field was rebuilt into the Baixiu Village Recreation and Sports Center Square. Capital investment for the project was said to be about 2.5 million yuan. The large, multipurpose square, housing sports and recreation facilities, is made up of a huge sports, library, and entertainment complex, situated next to a 30,000 square-meter cultural square. The indoor sports stadium has sixteen ping-pong tables and billiard tables. There is also an open-air sports ground with seating on three sides, which can be used as either a basketball or a badminton court. In addition, there is a grass football field and a swimming pool. The villagers told us that after work hours in the evenings, all the sports facilities are packed with workers from other provinces. The library houses about 20,000 volumes and is equipped with an electronic antitheft alarm system and a reading room that can seat over 100 people. According to the villagers, the library users are also from other provinces. The cinema, covering about 2,000 square meters, has 990 velvet seats. In addition to basic facilities like air-conditioning, it also has a Dolby surround-sound system and other advanced equipment. It presents up-to-date feature films from China and foreign countries. Tickets are inexpensive, costing only 5 to 8 yuan, so many workers from other provinces attend the movies. The cultural square is filled with many trees and plants, greatly improving the environment of the industrial area. It is a place for leisure activities and walking. There is an open-air stage that hosts many kinds of cultural performances and singing contests.

During the past six years, Baixiu Village has changed a great deal. Large-scale construction projects have given Baixiu a reputation and status far above that of nearby villages. Naturally, they provide the villagers with a sense of pride. As the taxi was about to turn into the industrial area, the driver, who is a Baixiu villager, quite proudly pointed out to us that the road had cost 7 million yuan. He obviously relished telling us that, drawing our attention to the two large characters "Bai" and "Xiao" conspicuously printed on a road sign.

In general, the facilities serve cultural and recreational purposes, and the patrons are mostly workers from other provinces. The Baixiu Market has the most direct effect on the daily lives of Baixiu villagers. The market is a large, covered structure in the industrial area. The roof of the building is shaped in a geometric pattern formed by the intersection of straight steel columns. From afar, it looks like an inverted ladder with a large rectangular protusion, resembling an ultra-modern sports stadium. The market occupies an area of about 16,000 square meters. There are more than fifty stalls arranged in neat rows. The rows are divided into zones, each selling different kinds of produce: dried foods, fresh meats, fish, vegetables, and so on. Bigger shops and small eateries line the four walls. One of the research team members compared this market to a large European marketplace. There are still about 2,000 local villagers in Baixiu Village. From the end of 1994 to 2001, the number of factories doubled, from over forty to over eighty, and there are about 40,000 workers. The new market in the industrial area meets the needs of this huge and rapidly increasing population of workers. It replaces the old open-air gathering of vendors that used to take place in an area in the midst of the living quarters. Now when the locals want to shop, they go to this market or to the market in neighboring Xijing Village.

Revisiting the Local Male Villagers: Wang Zhenqiu, Li Zhichao, and Wang Qiming

I first became acquainted with Wang Zhenqiu, Uncle Qiu, in 1994; he was then over sixty years old. He is a barber. When business is slack, he enjoys watching television, playing poker, chatting, or resting on a sofa at the Fraternity Association. In recent years, because of squabbles over the relocation of the ancestral tombs and other village affairs, Uncle Qiu has run afoul of some relatives in the Dezu branch of the Wang lineage. So he seldom goes to the association anymore. Instead, when there are few customers, he simply rests in his shop or goes home to take a nap.

The quarrel over the relocation of the ancestral tombs took place about four or five years ago. Some elders of the Dezu branch of the lineage felt that the fengshui of the tombs of the Dezu ancestors no longer had a beneficial influence to protect their descendants. In addition, the tombs were far away from the village. Therefore, it was proposed that the tombs be moved back to the village. Most young people of the Dezu branch accepted this proposal, whereas Uncle Qiu and some other elders did not. They felt that the proposal was only motivated out of material gain. The ancestral tombs might have lost their influence, but the ancestors had been buried there for over a hundred years. To move the tombs merely for one's own benefit was an act of profound disrespect. It revealed a lack of understanding about funeral rites and a lack of respect for ancestral worship. As to the argument that the tombs were too far away, Uncle Qiu claimed that in the old days, transportation was less convenient, but still the people would ride bicycles or walk to the tombs during the spring and autumn worship. Now there were cars and buses, but there were more complaints and fewer worshippers. Respect for ancestors had declined. The tug-of-war over the moving of the tombs continued for a few years. Finally, in 1997 the villagers decided to move the ancestral tombs back to Baixiu Village. During the course of this dispute, Uncle Qiu was one of the most vehement in opposing the move. As a result, he made some enemies among his male relatives in the Dezu branch.

One afternoon in the summer of 1999 I went to Baixiu Village and visited Uncle Qiu. He was not in his shop, so I went to his house. He was resting in a chair in the large sitting room. His family members had all moved, and he was left alone in the big, dilapidated house.

Uncle Qiu told me that two years earlier, his daughter-in-law and two granddaughters had received official approval to be reunited with his son in Hong Kong. He was pleased about that, as it meant his son would stay put. On the other hand, he was worried about the heavy financial load on his son.

"Now, it's one man working, five mouths to feed. It's tough," he said. Perhaps he worried too much; his face was sallow. "I heard that my daughter-in-law is looking for a job. If she gets one, no one will look after the two girls, and they may get into trouble. There are all sorts of characters living in their public housing complex."

"Surely Grandma can take care of them," I said.

"She's getting very old!" Uncle Qiu said. During these two years since the granddaughters moved to Hong Kong, he has had a good excuse to phone his wife. Unfortunately, she is unwilling to return to Baixiu Village. It is clear that he thinks of reuniting with his wife in the village.

When I revisited the folks in Baixiu Village at the beginning of 2002, Uncle Qiu was almost seventy. His body was much frailer than before, yet his spirits seemed high. While at a restaurant, he told me that he has cataracts and his eyesight had deteriorated.

"Because of my poor eyesight, I can now only give two or three haircuts each day. My income has dropped. Can you tell whether the left eye is worse than the right?" he said, opening a folder that contained his medical records.

I could not help but sympathize with him. "What happens when one's income decreases?" I asked.

"My son came back for a visit this year. Learning of my cataracts, he told me not to be worry. He said, 'Don't be afraid, Dad. If you have no income, I'll take care of you,'" Qiu said with an air of quiet relief. Perhaps this explained the reason for his high spirits.

Elders like Uncle Qiu cling to the ideal of a harmonious, intact family life and the traditional lineage ethics. The reforms and open-door policy over the past more than twenty years have definitely improved the material lives of the villagers. However, the reforms have undermined the old values and beliefs. In an atmosphere in which economic benefit prevails over other considerations, the traditionalists appear to be behind the times. They may even be ridiculed. The elders who opposed moving the tombs inevitably were defeated. Because he lives alone, Uncle Qiu is more lonely than other elders in the village. We do not know if his son will keep his promise to take care of him. After all, the younger generation is constantly being influenced by new concepts and values. Amid the vast rapid social changes that have engulfed the village, elders can only derive a sense of security from the love and care of their children. This is one of the main ingredients in traditional Chinese familial culture, hence the Chinese belief that "Father and son are intimately related."

When I got to know Li Zhichao in 1994, he was a middle-aged villager in Baixiu, around forty yeas old. In the past years, I had noted that Zhichao worked very hard to make money through all sorts of endeavors. He tried opening a courier service, but the company collapsed in less than a week. He then tried to partner with some others to run a lychee farm but it did not succeed either.

After the 1997 economic crisis in the Asian financial markets, the economy of the Pearl River Delta took a turn for the worse, and Baixiu Village was adversely impacted. According to Zhichao, since 1998, he was earning a little over 1,000 yuan per month from his motorcycle business, a drop of over 50 percent compared to the peak period. With the collapse of his courier service, his savings had been depleted. At the same time, the expenses for his son and his daughter who were now in senior high school increased.

"Two children in school means many more expenses. If Little Jun goes to university, my expenses will multiply even more," he said. Zhichao and Chao Sao pinned their hopes on their son Shijun. In 1999, Shijun took the university entrance examination but did not do well, and was not admitted to university. Extremely disappointed, he shut himself up in his room all day. Zhichao, and especially Chao Sao, were very worried about him. Chao Sao could not sleep or eat normally even a month after the exam results were released. I met her when I visited Zhichao on my return to the village, and I was shocked by her pallor and size—she had lost over ten pounds. Shijun had gone to Guangzhou to enroll in short courses and tutorial classes in the hope of retaking the entrance examination.

"Tutorial classes and certificate courses are costly," Zhichao told me. He was worried about his son's future and feeling the burden of the economic responsibility.

One afternoon in mid-2000, I took a walk with Uncle Qiu in the industrial area. I noticed a twenty-passenger bus. Uncle Qiu told me that such mid-sized buses now travel the roads between Baixiu Village and Qingyang Town and the nearby villages. The fare from Baixiu Village to the nearby villages is 1 yuan, and from Baixiu to Qingyang is 2 yuan. The buses are both inexpensive and convenient and are welcomed by commuters. After buses were allowed in Baixiu Village, the motorcycle business declined sharply. I broached the subject with Zhichao. He said that with the economic downturn since 1998 he had to work twice as hard as before, but he could earn over 2,000 yuan per month in a good month. But since the start of the bus service, his income had dropped to about 1,000 yuan per month, or even less.

"Little Jun is now studying at Dongguang Polytechnic. Wailing is in senior high. I figure that our expenses total 5,000 yuan a month. My income from driving the motorcycle will not be able to cover them all. Fortunately, I also have 2,000 yuan income from rent and my wife earns over 1,000 yuan. Otherwise we would be in the red." Zhichao had quit gambling in order to pay for his son's tuition. But the new buses had been unexpected. Moreover, he was beginning to feel the effects of age. He was less adept at vying with competitors for customers. Sometimes he even felt that he did not have enough physical strength. He knew that he could not go on like this, so he began to think again about other businesses.

Zhichao's relative Ah Kam suggested that he buy a car and operate an on-call taxi service. Zhichao could not make up his mind about it. In the Qingyang Town area a license would likely cost 150,000 yuan and be valid for only seven years. But in the villages and towns in Guangdong most taxi services operated without licenses because the licenses were so expensive.

"It would cost over 100,000 yuan to buy a car, and even more if it were a foreign car—maybe 300,000 yuan. I can get a loan from the bank for about 100,000 yuan using my house and land as collateral. Including bonuses, our family can expect 30,000 yuan this year. But what happens if business is not good after we invest 100,000 yuan? Also, now there are more and more people in the taxi business.

"My brother told me that if we give it our all, working sixteen hours a day, we can get a net income of over 10,000 yuan a month. But I'm getting on in years. Working day and night will tax my concentration and physical strength. What happens if there is an accident? Then I will lose everything including the investment?" In considering the pros and cons of going into the taxi business, Zhichao was no longer the gambler he had been a few years earlier. He was no longer so sure of himself in his decisions.

It was also hard to decide what kind of car to buy. "The best kind is the 'bullet car' [a well-known Japanese seven-seater]. It is wide-bodied and luxurious, and the most attractive to customers including both Hong Kong and Taiwanese bosses. There are two such cars in the industrial district. But they are too expensive! Over 400,000 yuan each. And China is about to join the World Trade Organization. After that, taxes on cars will decrease. Buying a new car now will cost more than buying one after China joins the World Trade Organization." In considering the investment, Zhichao was aware of the ways in which the world economy impinged on his decisions.

As pressure on the household economy mounted relentlessly, Zhichao finally bought a China-made jeep and started an on-call taxi service. Once, when I was riding in Zhichao's car, he said to me: "When I began my taxi service, I was nervous." He had already been in the business for over half a year when he told me that.

"What? Driving a car took a bit of getting used to, right?" I teased him.

"My driving skill was not the problem. When I drove the motorcycle in the summer, I wore long-sleeved shirts. But even so, my face and both hands would be tanned by the sun. When I changed jobs, I hardly had any money and I couldn't buy the proper clothing. A tanned guy in sloppy clothes! When I was driving on highways, it was easy for the police to mistake me for a thief. Whenever I was stopped and questioned, there would always be big trouble." Like many of the taxis in the village area, Zhichao's taxi did not have a license. If he were to be stopped by the traffic police and found to be operating illegally, the penalty could be as high as 10,000 yuan.

"I was stopped once by a traffic policeman, just one month into the new job, at a toll booth. I was so scared! Luckily, he was only checking to see if I had my driver's license."

"Were you ever stopped again after that incident?" I asked.

"No. I've been lucky! In the summer I had to drive the motorcycle in the rain and sun. In the winter, no matter how much I wore I always felt the chill in my bones. And I was always exhausted. Even if I could wear better clothes, they wouldn't suit me, and I wouldn't look like the chauffeur of a taxi." Thinking back to those days, Zhichao still felt a sense of inferiority. "Now it's much better. Inside a car, I have air-conditioning and heat—warm air in winter and cool air in summer."

It was obvious that Zhichao was better dressed now, and his skin was no longer as tan as before. "Now that you will no longer be mistaken for a thief, you must not be worried any more, right?"

After becoming a taxi driver, Zhichao had a better income. His son could continue studying at the polytechnic and his daughter could go to a post-secondary college. Zhichao felt that he had made the right choice.

"At present, excluding expenses on the car and repaying the monthly loan, I get over 4,000 yuan per month. When business was brisk one month, I earned 9,000 yuan! The job is much more comfortable than before. And I have become more competitive!"

"Why competitive?" I was puzzled.

"As a taxi driver, you have to know how to hook a customer. One day, a customer who knew me well wanted to charter the car for half a day to go to Zhongshan. Normally such a trip costs 500 yuan, with all tolls paid by the driver. So I told him, '500 yuan is the normal rate.' I did not get the deal. I found out later that another taxi driver had charged him only 450 yuan, and he went for it. I never saw that customer again!" Zhichao was obviously very unhappy about that.

He continued, "If a regular customer asks for you and you aren't available, you can pass him on to one of your relatives. You must never pass him on to other taxi drivers. Those drivers will steal your customers by lowering the rates. Under such circumstances, I too steal customers from other drivers by offering lower fares. For example, the fare from Baixiu Village to Qingyang Town is normally 30 yuan. Some workers from other provinces might offer 25 yuan. Hoping that they will come to me again, I will accept their 25 yuan. The golden rule is to keep your old customers and to steal new ones. That's the only way to survive."

The so-called "competitive" approach is not unique to Zhichao; other taxi drivers behave similarly. What they mean by being "competitive" is to grab customers by offering lower rates. It is cold-blooded because it shows a callous disregard for the previous agreement on pricing. Also, stealing another driver's customers by offering a lower rate seems not quite right. After all, all the taxi drivers in the industrial area are villagers from Baixiu Village. Some are related to one another by blood or marriage. However, they clearly know that if they do not act in this way, they will not be able to survive in this cut-throat business.

Since becoming a taxi driver, Zhichao had cut back on "playing with women." "My wife thought that once I became a taxi driver, then I would "play with women" regularly. The truth is I do that much less now. I probably do it less than once a month since changing jobs."

"You're kidding!" I was really surprised.

"When I first took this job, I just sat in the car even if there was no business. I was afraid that customers might call me on the phone in the car and I wouldn't be there to answer. When business is good, there's just no time for amusement." So he had not really changed. It was just that he was concentrating on his work.

"Around the middle of 2001, business was booming in the evening. Often, some bosses would ask me to take them to karaoke nightclubs, and even invite me to stay to enjoy myself and call a dance-girl as an escort. I really wanted to stay but if the boss paid for me, that would not be right. He was, after all, my customer. It's a bit too much asking him to pay," said Zhichao. He told me that instead he would go outside to chain smoke, not wanting to gawk.

"One day, some customers chartered my taxi and went to a nearby town to have sauna and massage. At first I wanted to wait for them in the car. But they were generous and insisted on my going in and paid for me. Of course, I paid the tips for 'playing with women.' It was a great experience!

"The girls now cost much less than in previous years. Too many of them have entered this trade. A few days ago, someone telephoned me and said he had found me a girl who had just joined the trade. Sometimes when the Hunanese want to get a job, they will voluntarily send some girls over. But driving a taxi is a full-time job, no time to fool around." Zhichao had cut down on the time spent on "playing with women," but that does not mean he had changed his views about women at all. For him, women were still "like soda pop cans and chewing gum. You drink it, you throw it away; you chew it, you spit it."

Zhichao is similar to other villagers of his generation in pursuing the market economy. They seek every opportunity to better their income. By participating in market transactions, they learn how to calculate profits, and, gradually, how to make profits under the rules of market competition. However, they are not profiteers. When there is a choice between making more profits and maintaining a harmonious personal relationship with lineage members, they would rather choose the latter. It is obvious that the constraints of the traditional familial culture still exert a deep influence on villagers like Zhichao. As in earlier generations, Zhichao is the hub of his family, holding unquestioned responsibility to be the keeper of the family and to take care of his wife and children. This can be said to be a positive consequence of the tradition of the male forming the axis of the family. (It is what is sometimes called a "patriarchal culture.") On the other hand, outside the family, such patriarchal dominance results in an unequal relationship between males and females, as is evident in Zhichao's attitudes toward women.

Wang Qiming is younger than Uncle Qiu and Zhichao. After his marriage, he still lived with his parents for financial reasons. He has always dreamed of having his own house. This would reduce the friction between his wife Meiling and his mother. At the beginning of 1997, I learned that Qiming finally had a house of his own. One evening, I went to his new house to pay him a visit. It is a two-storied building, situated beside a fish pond in the northwest side of Baixiu Village. The building looked quite plain from the outside. Inside, the furniture and decorations were modern and fashionable. Qiming told me that the house and furnishings had cost almost 400,000 yuan.

I remember in 1995 Qiming told me that he would not be able to save enough money in three or four years' time to build his own house. How did he amass enough in the short span of one year to build a house that cost several hundred thousand yuan? I was curious, but I refrained from asking.

Qiming is now the general manager of a factory. I rarely saw him after 1997. On the few occasions when I met him in the village, we would only exchange a few words of greeting. Several times, I wanted to chat with him in the evening, but he was accompanying guests to a karaoke club. Later, I learned some things about Qiming from his uncle. In 1998 Qiming was found to have embezzled about 400,000 yuan from his factory. When the matter was discovered, Qiming and Uncle Quan went to the party secretary to ask for help. The secretary finally settled the matter for them. At the end of the same year, there was an election for members of the Village Committee. Both Qiming and his father worked hard to canvass votes for the secretary. Eventually the secretary won the election. Qiming also actively took part in the election, running for the leader of the fifth group; he too won. The next year, that is, in September 1999, with the help of the secretary, Qiming became the general manager of a foreign-owned factory. Both his salary and fringe benefits were considerable.

At last, at the end of 1999, I had a chance to have a talk with Qiming. He told me that he had changed jobs and was working at a Japanese-owned chemicals factory. The new factory was much larger than the Taiwanese-owned factory where he had formerly worked. Its regulations were strict, and Qiming learned a lot from working there.

"In the Taiwanese factory we seldom bothered with regulations. If a worker asked for leave to return to his village, we'd see who he was and whether he was an obedient worker, before we made a decision. But this Japanese factory is different. Everything has to be done according to rules and regulations. If the regulation says that a worker has the right to ask for leave, then we have to approve it regardless of who is applying."

"If the regulations are so strict, how can the factory managers make profits?" I asked. I knew for a fact that some factory managers in Baixiu Village often made profits on the strength of their positions. For example, some of them secretly sold the factory waste for extra income. Or they billed the company for their night out at karaoke nightclubs as business entertainment.

Qiming explained to me what he had learned in the Japanese company about doing things according to regulations. "In our company, you just cannot do those things. There are clear and strict procedures for reimbursements. It is better to keep your feet firmly on the ground. Even if you use all sorts of ingenious methods to get a profit by illegal means, you will be discovered one day." Perhaps Qiming had learned his lesson after the last disaster, and had become more careful.

"I heard that you've been elected group leader [more or less equivalent to a production team leader in the people's commune era]. Does it involve a lot of work?"

"Not that busy. The village is now quite well developed and things are all regulated. The farmland in our brigade has nearly all been sold. Our main concern now is to attract more foreign investment to bolster the income of the brigade."

Not long after that conversation, I took Qiming out for dinner. In the restaurant, quite by accident, we met his relative, Ah Wai. We talked about the Village Committee election and how Qiming and Ah Wai could make take advantage of opportunities to do business.

"It is a good idea to use plastic materials to make windows. The end products are both pretty and strong," said Ah Wai. Their dreams knew no boundaries, as they contemplated going into window making.

"The waste produced by my factory can be transformed into plastic material, and I can get my hands on the waste. Trouble is, I don't know the technology." It seemed that Qiming was once again thinking of taking advantage of his position to make a bit of extra money.

My impression of Qiming after our two conversations was that he had matured, but he was even more eager to go into business to get rich. In June 2000, Qiming rented a shop at the Baixiu Market. He invested around 20,000 yuan to open a Hong Kong–style restaurant. The restaurant attracted many Hong Kong truck drivers, bosses of Hong Kong–invested factories, and employees. Business seemed to be booming. One evening, I went to Qiming's restaurant for coffee. He was there looking after the customers.

"Business has been hectic since the restaurant opened. I come here to take a look around before going to the factory to work. Then I leave my wife and her sister in charge of the restaurant. Once I get off work, I come here and stay till two o'clock in the morning. There's hardly any time to take a break."

Qiming felt that the work was tiring, but when he talked about the restaurant, he seemed to be full of enthusiasm.

After a while, Qiming's wife and two children came in. Qiming told me that when his wife came to work in the evenings, she always brought the children. Family life, such as it was, was spent in the restaurant. According to what I could gather, during these few years, Qiming no longer frequented the karaoke nightclubs to "play with women." There were only occasional visits, sometimes because of business, but most of the time he stayed at home to take care of his family.

The young villagers of Qiming's generation are even more intent than those of Zhichao's generation to fulfill their ambitions and advance their careers. At the same time, they also pay more attention to marital feelings. The familial relationship has slowly become centered around the husband-wife axis.[1] Since this axis excludes outsiders, it may partly explain why Qiming now seldom frequents sex spots to "play with women."

In the past seven to eight years, the sex trade in Qingyang Town and Baixiu Village, such as karaoke nightclubs and hair salons, has been reduced by 60 to 70 percent. Now the favorite pastime is to go to discothèques, which are modeled after those in Hong Kong and Shenzhen, where there are escort ladies with whom one can sit, drink, or dance. Of course sexual services are still available. What is new is that these places attract a younger clientele who dance and drink. They often take soft drugs such as ecstasy, which has become quite popular in the Pearl River Delta area.

In the 1990s and up to 2002, there was a steep decline in the sex trade businesses. One reason for this is that the outer economic zones, such as Hong Kong and Taiwan, had suffered an economic downturn for some years, which affected the spending power of people in the area. Second, the sex trade was emerging and prospering in the interior provinces. Women from interior provinces no longer were pouring into the Pearl River Delta. After more than a decade of growth, the sex trade was not as attractive as it once was. This is not to say that male–female relationships have not suffered from indiscretions. But from our observations, sexual relationships between men and women quite often now take place in factories.

Revisiting the Female Local Villagers: Li Shimu, Wang Guizhen, and Dayan Sao

I became acquainted with Li Shimu during the winter of 1994. She was then sixty-nine years old, a healthy and talkative woman. In 1996, she came to Hong Kong to visit her youngest son and we met again. Still healthy, she loved to travel. Five years earlier, Li Shimu had told me that her children all had their own families and the grandchildren were healthy. She could leave the world without any regrets. Born in the twenties, Li Shimu had seen both war and turmoil. In 1949 when the new government was formed, she followed her husband back to Baixiu Village where she spent the next half century. Her husband was a teacher. He died in 1974, toward the end of the Cultural Revolution. I do not know, and never asked, how her husband died. What I do know is that after many decades of ups and downs, campaign after campaign, Li Shimu has acquired a come-what-may attitude toward life.

One afternoon in the summer of 2000 I went to Baixiu Village to visit Li Shimu. She was then seventy-five. I had heard that she had had a new house built. A kind old woman directed me past a few turns and led me to Li Shimu's new house. I looked up to find a silvery iron gate at the door, behind which was a concrete two-story building with a flat roof. It stood out from the old buildings with red bricks and upturned ribbed eaves. Her house was built in the new style of architecture that was so fashionable in Baixiu Village.

Li Shimu was not at home, but her grandchild, Ah Wen, whom I had tutored in English, led me into the house. It was somewhat untidy—a few cups, some peanuts, paper, and other things strewn on the tea table. I remembered that Li Shimu did not have a tea table in her former house. In the sitting room there had been a wooden bench and a table that served as a dinner table, desk, and mahjong table. The sitting room in the new house was spacious, about double the size of the one in the old house. In addition to the tea table, there was a sofa, canvas chair, and some other modern furniture. It was clear that Li Shimu had become affluent, and like many villagers in Baixiu Village aspired to improve her living environment.

I asked Ah Wen whether Li Shimu had gone to play mahjong. Ah Wen told me that her grandmother seldom played these days. Instead, she had gone to help out at the eatery in the "big market." The big market was the newly constructed Baixiu Market, open for business only since June. Li Shimu and her family had successfully bid on a unit in the market that they turned into an eatery. In the morning they served breakfast, rice rolls, congee, and rice noodles. At noon and in the early and late evening, they served fried noodles. The shop was managed by Ah Hui, Ah Wen's elder brother. Li Shimu usually went to the market to buy the vegetables and meat and then she helped out in the shop.

The eatery was clearly an investment by the whole family. Ah Wen told me that her father had changed jobs and now was a motorcycle driver. When he was not gambling, he would go to the eatery to help out with various jobs. As for Ah Wen, he was waiting for publication of the results of the university entrance examination due the next month, so he also helped out at the eatery.

I changed the topic to the new house. Ah Wen said that it had been built two years earlier. The top floor was for his elder brother; the first floor had two rooms—one for his parents and one for himself. Then he pointed to a spacious room with an open door on the ground floor. There was a large bed stacked with many bundles.

"This room was originally given to grandma, but she does not live here," said Ah Wen.

"Why?"

"She keeps saying that someone has to live in the old house. If no one lives there, the beams will turn bad. She is not willing to rent the house to people from another province, so she stays there. In the daytime, she comes to our new house, and after dinner when it is bedtime, she returns to the old house."

I chatted with Ah Wen for a while. Before leaving, I asked him how Li Shimu was doing. He thought a little before answering "Good." I told him that I would go to their eatery to find Li Shimu the next morning.

The clock said four o'clock when I left Ah Wen. It was early and I knew that Ah Zhen (Wang Guizhen) and her husband had gone to Hongye Village. The reason has to do with clearing customs. They would not return until six o'clock in the evening. So in the meanwhile, I would make use of my time by visiting Ah Zhen's old house. I might find Nianzi and Fenkai.

After completing the fieldwork in Baixiu Village in 1995, I had an opportunity to meet the three sisters Ah Zhen, Nianzi, and Fenkai in Hong Kong in 1998 when they joined a tour to Hong Kong. When they came to see me, Ah Zhen told me that she had gotten married. Her husband, a Baixiu local, had returned from working in another town. I was happy to hear this news, and I could see that Ah Zhen had put on weight. Nianzi and Fenkai joked that she did not have to work and her in-laws did not expect her to do any housework. She only ate and slept. I could see that Nianzi and Fenkai, though making fun of their sister, were also a little jealous. Nowadays married women of Ah Zhen's generation do not have to help their husbands cultivating the rice fields nor do they have to do all the household chores, or look after their in-laws. Still, it is rare that one does not have to do any housework at all and has the meals prepared by a mother-in-law or a sister-in-law.

Leaving Li Shimu's house, I headed straight to Ah Zhen's old house. Now only Nianzi, Fenkai, Deming (the youngest brother), and Old Uncle Wang live there. I had heard that Deming was teaching in Qingyang Town and lived in the staff quarters and seldom came home. In effect, there were only three people living in the house—the two unmarried girls and their father.

Smiling, Fenkai opened the door for me and led me into the sitting room. I sat on the long wooden chair where Ah Zhen used to sit when she watched television. The furniture looked just as it had five years earlier, only older. Some of the paint on the walls was peeling. The dilapidated red bricks on the floor were chipped in two places and were disintegrating, exposing the sand and mud beneath and showing their age.

I asked Fenkai about the bricks. She looked annoyed and said that her father scrimped to a fault, and was not willing to spend money on renovations. Fenkai did not appear to be embarrassed to be living in such an old and run-down house. The reason could be that since then the house belongs to her father, and its condition is not a concern of the daughters. Traditionally, daughters belong to "someone else's family." Since they will not stay at the house forever, they do not really care about the appearance of the house. I knew that Fenkai, with her regular and part-time jobs, made the most of the three sisters' wages; before her marriage Ah Zhen came second, and Nianzi did not do badly as a salesgirl. If they had wanted, the three of them could have afforded to make some repairs. When they visited Hong Kong in 1998, they went on a spending spree in the shopping mall. I was surprised at how much they spent. The dilapidated house .was representative of a family with daughters and no sons; there was an unarticulated sense of a family without a future.[2]

I chatted with Fenkai for awhile, then the telephone rang. It was Ah Zhen. She said she was having a snack with her husband and Nianzi in Qingyang Town. She would return to the house after the snack.

I stayed on to talk with Fenkai and we turned to the topic of Ah Zhen's present situtation. Fenkai told me that Ah Zhen had given birth to a baby boy on September 1 of the past year. Her husband and in-laws doted on the baby. Fenkai and Nianzi also would occasionally go to Ah Zhen's house to play with him. The two houses are close by in the same village. On most days Ah Zhen would take the baby to her old house after dinner and chat with her sisters. It is clear that the three sisters have maintained their close relationship even since the eldest sister's marriage.

Soon I heard the sound of the iron gates opening outside. Ah Zhen, Nianzi, and a strong, well-tanned man came in. I thought he must be Ah Pong, Ah Zhen's husband. They sat down and Ah Pong introduced himself. Ah Zhen took out two big bags of lychees and longans and placed them on the tea table. As we ate, we chattered away on a variety of topics. There was the sound of laughter and a relaxed warmth emanating from our circle. Then they decided to go to Qingyang Town that evening to go bowling.

Ah Zhen invited me to join her for dinner at her in-laws' home. Fenkai and Nianzi would have dinner at their own house, but they would go with us to Ah Zhen's place to see their nephew. The five of us set out together, divided into two groups on two motorcycles. Fenkai and Ah Pong were the drivers. They weaved smoothly along the alleys between the houses in the village, reaching Ah Pong's house in less than two minutes.

Entering the front gate, we saw Ah Zhen's mother-in-law bathing her grandson in the front yard. The baby boy, with a lovely face, looked like Ah Zhen. Fenkai and Nianzi squatted alongside the bathtub to caress their little nephew. Ah Zhen's father-in-law, a tall and thin old man, was cooking dinner. Ah Pong led me into a spacious sitting room. The furniture and decorations were not extravagant. However, they were neat, shiningly clean, and comfortable. This contrasted sharply with the darkness and state of disrepair in the sisters' old house. Ah Pong told me that this house had been built in 1995 when they relocated to Baixiu from Shaoguan. The building was carefully designed. On the ground floor there was a sitting room, kitchen, and a room for the parents. The first floor had been designated as Ah Pong's living quarters for after he married. Half of the second floor was reserved for a relative.

After dinner, Ah Pong showed me to the upper floors. The rooms were neat and tidy. From the top floor, wide on all sides, one could see the many ribbed rooftops on the old-style village houses interspersed with new flat roofs. In the distance was the brightly lit Baixiu Market.

I asked Ah Pong about his decision to return to Baixiu and his adaptation to life here. He was not a chatterbox, but he was quite ready to talk about his own affairs and his feelings.

"I lived in Shaoguan for twenty-two years. Most of my classmates lived there. I worked as a welder; it's a great steady job! But because my dad made up his mind to return to Baixiu, then return it was. I moved back here in 1995."

Although Ah Pong was a local, he said he felt like a foreigner. I asked him about his life here over the past five years.

Looking at the faraway rooftops, he did not answer me directly. "At first I was like a fish out of water. Here I have only relatives, no friends." Because it was dark, I could only make out his silhouette. "It is more modern here. When I lived in Shaoguan, I lived in the countryside. The people there were simple and friendly. They would talk and laugh with you. I have a lot of friends there."

"Are your acquaintances here mainly your relatives?" I asked.

"Maybe it is because the people here are rich. You feel like they look down on you." Ah Pong seemed to be lost in his thoughts. After a silence he continued, "When I first got back, I was not used to the water, the food, or the environment. Now it is better."

At ten o'clock, Fenkai and Nianzi returned from dinner at their own house, and we made our way into Qingyang Town. Ah Zhen brought her baby along. The motorcycles turned onto the highway after leaving Baixiu Village. But in no time, we left the highway and swerved onto the main road of Qingyang Town. Several twists and turns later, we reached the bowling alley. I glanced at the clock; it was forty minutes past ten. There were many people inside. Fenkai went to the information counter and returned to tell us that we would have to wait for some lanes to clear. I was somewhat surprised, not expecting the place to be so busy. It could be because the rates were less expensive late at night only 7 yuan per game. A while later, we got a lane. Fenkai and Ah Pong were the most skilled bowlers; Ah Zhen was about my level. Nianzi did not bowl but sat there looking after her little nephew. We played a total of twelve games before we paid the bill and left.

Emerging from the bowling alley, Fenkai suggested we go to Fuheng Restaurant to have a late-night tea. I stole a glance at my watch, which said nearly half past midnight. In fact, I was quite tired, but seeing that they were all in high spirits, I forced myself to wake up so as not to be a killjoy.

After the midnight snack, past one o'clock, we returned to Baixiu Village. We passed many shops that were still open, some of them restaurants, some of them clothing shops. I was surprised that commercial activities continued so late into the night in Qingyang Town.

The next morning, I went to the Baixiu Market to buy rice rolls. Ah Wen and his father Ruqing were already working. Ruqing, with a crewcut, looked younger than five years earlier.

Ruqing told me that his mother would come out to the shop only after finishing the household chores, so she would be there later in the morning. I looked around the small shop that was about 100 square feet. It was sparsely furnished with some rectangular tables. In the outermost part of the shop there was a big cooking stove for making rice rolls. In the innermost part there was a heavy wooden round table for chopping barbecued pork. Ah Wen's brother Ah Hui was busy as cashier. Ah Wen told me that they had hired two workers from other provinces, one to make rice rolls and the other to do other work. Their combined wages were 900 yuan. Ruqing noted that the interior decorations had cost them 20,000 yuan.

We conversed a bit about business. Then I asked Ruqing, "How is Shimu's health these days?"

"She was hospitalized two or three years ago. It was a serious illness. She almost lost her life."

I gave a sigh of relief, glad to hear that Li Shimu had made it through. Not long thereafter, Ruqing looked in the direction of the main entrance to the market and said, "Here she comes."

About twenty feet away, I saw Li Shimu coming toward us carrying a handbag. She walked unsteadily, swaying to the left and right. Ruqing pointed out to me, "She is shorter than before because her backbone is now curved to one side."

Five years had elapsed since we had last met, and Li Shimu was clearly thinner than before. Her face bore a worn expression. This was probably due to her inability to recover fully from surgery. After all, she was quite old.

Li Shimu greeted me with happy recognition. After sitting down we talked casually. She repeated her thanks to me for giving Ah Wen English lessons. And she told me that soon the results of the examination would be published and she would know whether her grandson could enter the university. In this regard, Li Shimu had not changed a bit: she was very concerned about her grandson's studies. We talked for about twenty minutes until I had to leave. She kindly told Ruqing to take me back to my hotel on his motorcycle.

During the visit in the summer of 2000, I had managed to contact Wang Guizhen and Li Shimu, two of the local women I was quite familiar with. But I could not find Dayan Sao. It was said that she had left for Hong Kong two years earlier. Although I was disappointed, I was happy for her. Dayan Sao, like many other women in the village, had always hoped to emigrate to Hong Kong to be reunited with her husband. At the same time, I knew that after emigrating, there are many adjustment problems. After long separations, there are often marital conflicts, sometimes leading to divorce. I wondered how Dayan Sao had survived these two years in Hong Kong. Unfortunately, I had lost all contact with her. I can only hope that she has been lucky. "Follow one's father when young; follow one's husband when married; follow one's son when old." That seems to be the fate of Chinese women in history, But nowadays, due to the complexity of circumstances, things have changed for women. Still, in the hearts of many Chinese women, a "blessed" life is one that conforms with this old saying.

Female Workers from Outside Provinces: An Update on Wang Qingmei and Li Mei

In April 2000, I called Wang Qingmei and told her that I would be visiting Baixiu Village. I made an appointment to see both her and Li Mei. Wang told me that Li Mei was now working in a factory in Shenzhen. She could come back to Baixiu Village during the weekend to see me.

It was a Saturday evening. At half past five I went to the entrance of the Xingsheng Electronics Factory to wait for Wang Qingmei. The guard inquired about my identity, then took me past the main door. He pointed to a long bench on a grass patch where I could sit down to wait. He was very courteous; evidently, Wang Qingmei and her husband had left word with him to watch for me. It was obvious that they were workers of some importance in the factory. Wang Qingmei had joined the factory in 1992, and had been working there for eight years.

Then Wang Qingmei appeared. She was not much different from before, only thinner. Her long hair was bound by a rubber band to form a ponytail. Typically, she was wearing a worker's uniform on the top, with jeans on the bottom. Like her old self, she welcomed me with a broad smile.

Wang Qingmei had gotten married and had had a son. When the baby was six months old, her mother took him back to Sichuan to take care of him. Wang Qingmei now lived in the married quarters of the factory dormitory assigned to her. She told me that the quarters were only for couples who both worked in the factory. I saw quite a number of children around the factory, mostly one to two years of age. Wang Qingmei said that many young couples had had children in the past few years.

Li Mei arrived at seven o'clock in the evening. She was visibly thinner. Equally visible was her love of cosmetics, and she knew how to use them. Her lips were colored deep red and her hair was dyed a fashionable brownish blonde.

I invited Wang Qingmei and Li Mei to come to the Huanan Hotel after dinner to spend the night so we could talk. When we arrived at the hotel, the television in the room had no antennae and there were only snowflakes on the screen. Without the distraction of the television, I had a good opportunity for a long chat with them.

Wang Qingmei told her story first. She was now working at the best factory in Baixiu Village. She only had to work two and a half hours of overtime, and Sundays were usually days of rest. In the factory there were workers who were university students, post-secondary students, and many secondary technical students. Not long ago, the company was listed on Hong Kong's stock exchange. The price was 1 yuan per share. Party members in the company could buy shares but they could not take them with them if they left. If employees sold their shares, they could not buy again. Wang Qingmei and her husband had bought 8,000 shares. She told me that the shares had risen to 3.50 yuan each. Many employees had sold their shares. She was thinking of selling 4,000 shares and leaving the rest as a long-term investment.

I asked Wang Qingmei: "After so many years of working, how have your ideas changed?" She said, with a sense of profound understanding, "I have learned how important money is." Wang Qingmei said that her ambition was to buy a house, but that after so many years of working, she still could not afford one. If she had a lot of money, she could bring her child from Sichuan and then the whole family would be together. She wanted to buy a house in the county town. She liked the type of houses where she would have her own kitchen and bathroom. Although houses in the county town were more expensive, she thought it was worth it because living in the county town was a status symbol. It represented a leap in upward mobility, and the children would receive a better education. Wang Qingmei said that she was careful with money and did not spend it on snacks.

I asked Wang Qingmei and Li Mei whether the phenomenon of "mistresses" in the factory was still prevalent. "Not so widespread now. On the other hand, I cannot say it is definitely not widespread. Maybe there are fewer of them in our factory. A rabbit does not munch on the grass near its warren. The people at the top can always find mistresses outside of the factory. It's all hush-hush."

"Is there someone keeping an eye on such things?"

Wang Qingmei shook her head. "But if exposed someone will get a bad name."

Li Mei had been away from Baixiu Village for a long time, so she did not know the situation well. As a result, she was quiet, lying on one side and listening to our conversation. I turned to ask her what changes she had noticed in her own thinking after so many years of working. Would she consider getting married? Li Mei said, lethargically, "I don't know." Then, "I don't want to get married. The marriages I see are not happy ones. The couples are always quarreling. I don't see the point of marrying."

Wang Qingmei said, in all earnestness, "All men are fickle." Li Mei told me that when she first came to Baixiu to work, a man from her village pursued her. He had come to the factory before her. When Li was still in the village she had heard that he was doing quite well, earning several thousand yuan a month. She guessed that she would have a good life if she were to marry him. In her village, there were few divorces. When people married, their fates were sealed. No matter what happened, they would carry on as husband and wife. Now of course she no longer agreed with that. She sighed, "Good men are hard to find. For people of my age, the good ones have all been taken."

I changed the topic and asked her when she had changed to her new job. She told me that when she came to Baixiu in 1994, she first worked in the Xingsheng Electronics Factory for three years. Then she was sidelined, so in 1997, she got a job in Xinke factory near Qingyang Town. Before going to Xinke, Li bought a post-graduate diploma without studying for it. Armed with the diploma, she got a new job, a better one, earning over 2,000 yuan a month. In 1999 Xinke was sold to another company. Then, through a person she knew, she moved to a Jiaxing factory in Shenzhen. At first, her contact recommended her to be in charge of a section, but her English was too poor to do the job. She ended up as a senior clerk, and she was not very happy. She said that with the fake diploma, she felt herself equal to the others in Xinke. She thought quite well of herself. But now in Shenzhen, things were different. People with academic qualifications were a dime a dozen. She felt a sense of inferiority among people more capable than herself. This extended even to her clothing. If she did not dress well compared to others, she would be upset. She wanted very much to further her studies. She wanted to learn English and get a specialized diploma.

Since that meeting in April 2000, I have kept in touch with Wang Qingmei and Li Mei by phone. Occasionally I get an e-mail from Wang Qingmei. At the beginning of the next year, Wang Qingmei phoned and told me that Li Mei's father had passed away a while earlier. Sometime after that, I heard that Li Mei had left Guangdong to go to Tibet to work. It seemed that she wanted to go west to explore chances for development. But, six months later, in November, Li Mei returned to Guangdong. She threw herself into the task of finding a job.

The three of us met up again on Sunday, January 6, 2002. At eleven forty in the morning, I waited for them on the ground floor of Wang Qingmei's dormitory in the Baixiu industrial area.

After the lapse of a year, Wang Qingmei did not look much different. Her hair was cut short and as a result she seemed to be more mature. Li Mei was busy and would arrive a little later. Wang Qingmei said that Li Mei had not been happy in Tibet. Luckily, she got a new job soon after her return. She was now working at a real estate company in Zhangmutou Town.

During my last visit to Wang Qingmei, she and her husband were sharing a big room with another couple. The bathroom was also shared. In the last year, Wang Qingmei had been upgraded to Grade 3, and her husband to Grade 2, giving them a combined grade of 5. So in July they were allotted a larger living area. Then they needed more furniture. In addition to the old iron bunk bed, desk, television, and refrigerator, there was a single bed and a water dispenser. When they shared a room, they had had to do the cooking in their cubicle. Now they had their own bathroom, where they put the cooker. Compared to their previous living situation, and the situation of most workers form other provinces, theirs was nothing to be ashamed of.

Wang Qingmei courteously asked me to sit down and gave me a few photo albums to look at while she prepared lunch. As I turned the pages, the first photo that caught my eye was a picture of her son Little Fangsheng. I had seen photos of Little Fangsheng before and I had vague memories of a little baby. Now, he was already a big child, in fact three years old. I remarked how lovable he looked. Wang Qingmei was clearly proud of her son. She said that Li Mei was jealous of her because Wang had a home she could call her own.

Li Mei had still not arrived when I finished looking at the photos, so I used Wang Qingmei's cell phone to give her a call. The couple now had a cell phone of their own. Every time I visited Baixiu Village, I saw that their life had improved a little. Wang Qingmei said that many workers in the factory had cell phones, and about one-third of the clerical staff had them as well. She had bought her counterfeit phone in Hong Kong for only a few hundred yuan.

At almost twelve o'clock, Li Mei arrived in a rush. Even though she had been in Tibet for half a year, she had not changed, and still was fond of beautifying herself. Her hair was dyed a light brownish blonde, she was wearing a black clingy one-piece dress. She was her same boisterous self as well, trying to keep herself busy by helping. But she was more mature than the previous year; both in terms of manners and speech, she seemed to be less headstrong. Li Mei was working at Antai real estate company. Her boss was a Hong Kong woman over fifty years old. Li Mei worked as the secretary to the director, at a monthly salary of 2,000 yuan.

After dinner, the three of us took a walk. Since it was Sunday, many workers were strolling in the square of the industrial area. We went to a pavilion to sit down. Eating the orange we had brought, we began chatting. Wang Qingmei told us that she had gone to an employment agency in Dongguan city because she wanted to change jobs. I asked her why. She sighed and said sadly, "Well, I'm over thirty. I don't want to continue on like this. . . ."

I had always thought that Wang Qingmei and her husband were content with their present jobs. I knew that although they had been in Baixiu for quite some time, they were outsiders and would have great difficulty merging into the local society. But they had their own circle of acquaintants and friends, and they both had relatively stable jobs. And now? Did they want to return to Sichuan to do business? While I was wondering, Wang Qingmei revealed that she had just had a row with the section head. She angrily related the incident to us. Not that long ago, during a meeting in the office, the section head had reproached her in front of all her colleagues for not doing her job properly. She responded immediately, and there was a scene. Later on, the head took her aside and said that he was dissatisfied with a lot of other people as well, not just her. He had singled her out in front of others because she was a "veteran." Wang Qingmei was not convinced. She retorted, "As the longest-serving 'veteran,' you should *not* have made a fool of me." She huffed, "I was here when the section was established. What's more, I am the best on the computer. If I had received a diploma, he would not have been the boss. I would have been the one in charge!" After that incident, she and her husband started thinking of leaving. The problem was that neither of them had diplomas. Wang Qingmei had taken the college entrance examination three times, all without success.

After listening to her story, I tried to calm her down and said she should not resign just because of this incident. I knew that jobs were not easy to find. A better strategy would be to try to get a diploma first before making a move. Li Mei had been quiet up to then but she broke her silence, "That is exactly what I told her. She would not listen. Now we have a second opinion." Wang Qingmei was silent, and seemed to be lost in some renewed thinking.

Both Li Mei and Wang Qingmei had left their Hubei homes to work in the enterprises in the villages and towns of the Pearl River Delta. Over the years, each had followed her own career path. Li Mei had been to Qingyang Town, Shenzhen, Tibet, and Zhangmutou to look for opportunities. Wang Qingmei had stayed at the Xingsheng Electronics Factory in Baixiu Village all these years, married, and had a child. Because of the open-door policy and economic reform that took place over the past decades, hundreds of thousands of female workers from other provinces had poured into the area. Li Mei's and Wang Qingmei's stories reflect the trajectories of some of the workers—how they went from some backward poor inland villages to the modernizing villages and towns. These village girls have undergone tremendous changes, in their style of dress, way of thinking, behavior, and outlook on life. What is common among them is the pursuit of better material conditions and the desire to attain some form of personal fulfillment. These common goals are proof of the development opportunities provided by the market economy. After marrying, these female workers prefer not to return to their backward villages if they can help it. The incentives of the market economy have lured village girls to leave their old ways in backward villages.

In conclusion, the lives of the people in Baixiu Village in the late twentieth and early twenty-first centuries have improved a great deal, drawing them closer to Hong Kong standards. Construction in the area, such as the networks of highways and roads, the Baixiu Market, schools, cultural and recreational facilities, sports arenas, and amusement facilities, are all up to standard and sizable. Most villagers have their own houses. Quite a few have their own cars. Trips to other places for vacation have become yearly events. Village life is no longer the erstwhile rural life. Other villages in the Pearl River Delta are moving in the same direction as well.

Our opinion, at the beginning of our research, that economic forces are far more influential than political forces, was confirmed. The period from the 1970s to the end of the 1980s can be seen as the first stage in the opening up of China. During this first period, only the Pearl River Delta area developed a market economy. Other places remained dormant. The period from the end of the 1980s to the middle of the 1990s constituted a second stage. During this time, the economic development of the Pearl River Delta reached a peak. Other areas in the province gradually caught up with the Pearl River Delta in terms of rate of development. The mid-1990s to the early twenty-first century witnessed the third stage. Not only Guangdong but all other provinces developed a market economy.

In other words, the market economy and the way of life it engenders have become an accepted and unquestioned way of life for the Chinese. Thus, when we look at the living conditions of the Chinese people at the turn of the millennium, we find that although they are still different from those in other developing or developed countries, the difference is only one of degree, not of kind. Although the poor interior provinces are far less developed than the Pearl River Delta, even they are far richer than before. Most families own a telephone, a television set, a refrigerator, and other modern family amenities.

To return to Baixiu Village, the difference between Baixiu and Hong Kong has narrowed much in terms of community construction and lifestyle. This is not to deny that at an ideological level, the people of Baixiu Village and those in free societies are still wide apart in terms of values—such as citizens' rights, equality between the sexes, and civic-mindedness. This seems to indicate a deliberate effort on the part of the ruling class in China to place restrictions on the political and even religious rights of its people. But, as emphasized repeatedly in this book, the march of the market economy and modernization cannot be blocked. Sooner or later, the market economy will inevitably lead to competition at the level of values. Economic development will bring with it a rising sense of civic consciousness and civil rights. Enhanced demands will manifest themselves first in cries for democratic elections.

This brings us to the theme of the next chapter, where we discuss political reform at the grassroots level in Baixiu Village. Our emphasis is on human relationships in this reform.

16

Rural Political Reform

The Baixiu Village Committee Election

China's reform and opening-up policy began in 1979. After more than twenty years, giant strides had been made in all economic and social aspects. The changes in Baixiu Village faithfully mirrored this epoch of fast-paced development. We finished our fieldwork in 1995, and completed the book in 1998. However, we still kept up our research by returning to Baixiu Village periodically. At the start of the twenty-first century, Baixiu Village has undergone tremendous advances in infrastructural construction. Economic and social developments are no longer so surprising. The village displays most of the characteristics of a modern society. The villagers' perception of life and other views are basically not much different from those of people in Chinese communities in capitalist societies like Hong Kong and Taiwan. Other than the appellation "socialist country," there are few signs of a socialist or communist regime in Baixiu.

Recent developments in China have been interpreted in a number of ways. One view is that reforms in the political sphere lag far behind the economic reforms. Some even suggest that China's economic reforms focus on only one thing—monetary profit—and there are no corresponding reforms in cultural endeavors. In particular, with the bankruptcy of the communist ideology, the people's spiritual lives seem to lack a sense of direction. The growth of a sense of individual autonomy is restricted to the sphere of economic self-interest. But from what we have observed in Baixiu Village, we find the situation to be far more complex, as we shall attempt to explain in this chapter.

First, the development of villager autonomy in Baixiu Village is not restricted to the economic sphere; it has deeper implications. Some of these were discussed in the previous chapter. In this chapter, we will examine the development of villager autonomy in the area of grassroots political reform.

In December 1998 Baixiu Village held its first-ever one-person-one-vote direct election to elect representatives to the Baixiu Village Assembly, the chairmen of six village groups (each group is equivalent to a production team in the people's commune era), and the leader, deputy leader, and three members of the Village Committee. As the first such election in the history of the PRC, even though there were various problems, the villagers participated enthusiastically. The problems included the following. The Baixiu Village Election Committee did not follow all the regulations to elect the village leaders, and some candidates offered bribes to get votes. Still, over 1,500 villagers voted, making a turnout rate of over 95 percent.

The work of the elected Village Committee consists first of instituting measures to implement state policies. It must follow stipulated rules to "ensure the implementation" of the major policies of the central government, such as birth control, tax collection, levying the agricultural tax paid in grain, environmental protection, and so on. Second, it also is involved in many undertakings related to the village—public affairs over the economy, politics, and culture of the village. The Committee is supposed to deal with all such affairs by "democratic decision making" through "democratic management," under the critical "democratic supervision" of the Village Assembly.[1] The town governments have no say in the affairs of the village; their only role is to provide guidance, support, and assistance.[2] In other words, it is the elected Village Committee that has real executive power to deal with the public and social affairs of the village.

The fact that the Chinese government has instituted Village Committees throughout all the villages in the country is seen by many as the beginning of democracy in rural society. Because villagers have been able to participate in one-person-one-vote elections to elect members of the Village Committee, they have gradually become familiar with concepts associated with democracy. As a result, the basis for democratic development has gradually broadened.[3] Other researchers point out that even though the election of the Village Committee has provided a chance for rural society to move in the direction of democratization, the villagers are still not familiar with the concepts of individual rights and individual autonomy. Thus it will not be easy for them to acquire the concept of political citizenship.[4] Take the case of Baixiu Village as an example. The villagers elected the chairman and other members of the Village Committee using the one-person-one-vote method, with a participation rate of over 95 percent of the eligible voters. In actual fact, to a large extent, however, voting was influenced by factors such as lineage. This means the Baixiu voters have not demonstrated the individual autonomy of Western democratic societies. No doubt, some scholars attribute this to the present Chinese political system, which has not undergone any fundamental reform. The Communist Party is still the only ruling party.[5] We should note that our main interest is not Chinese politics or the reform of its political system; rather it is to assess the development of the concept of individual autonomy among the Chinese, in particular among the inhabitants of Baixiu Village, through the election of the members of the Village Committee. As some have pointed out, traditional Chinese culture poses an obstacle to the growth of a sense of individual rights and individual autonomy and similar ideas.[6] In the remainder of this chapter, we shall approach this claim from three perspectives: familial culture, the individual's economic self-interest, and autonomy. We shall thus chart the development of the self-concept of the Chinese in contemporary society by focusing on the concept of autonomy in this ground-breaking election.

The PRC Village Committee Organic Law

As early as June 1988 the Chinese government began to employ communications and education to organize the grassroots administrative mechanisms of the rural areas into work guidance teams to promulgate the PRC Village Committee Organic Law (Trial), which was passed in 1987. The Ministry of Civil Affairs chose "demonstration units" in various provinces to implement Village Committee elections in the hope of extending them in the future to other places. According to reports, by the end of 1995 there were about 80,000 villages that had elected Village Committees, some as many as three times.[7] On November 4, 1998, the fifth meeting of the Standing Committee of the Ninth National People's Congress revised and passed the PRC Village Committee Organic Law, thus replacing the trial implementation law. It was planned that by the year 2000 all villages nationwide would have directly elected their Village Committees, and that all villages would be self-governed.

After the Third Plenary Session of the Eleventh Central Committee of the Chinese Communist Party in 1978, the household agricultural responsibility system of linking remuneration to output was implemented in all villages in China. The people's communes and their production brigades were abolished, as well as their control over the rights to the use of the villages' resources and to production and trading. These reforms helped to phase out the widespread poverty. In villages along the coast, especially in the Pearl River Delta, incomes increased substantially. This economic reform in the villages also brought about a series of changes in the systems of governance at the city, town, and village levels.

After the implementation of the household agricultural responsibility system in villages in 1979, the system of collectivization gradually crumbled. The people's communes, brigades, and production teams no longer held substantive power. As a result, village society fell into a state of anarchy. In about 1980, some villages formed self-governing structures to reduce the confusion brought about by the economic reforms.[8] In 1982, the Chinese government promulgated a new constitution in which the villagers' self-initiated structures were unified to become the Village Committee. The administrative tasks previously performed by the production brigades were transferred to the township governments. The day-to-day public affairs and production work were managed by the Village Committee, thus setting into motion the system of small-town-governing-villages. By 1986, there were about 70,000 township and town governments, and about 800,000 Village Committees. In 1987, the Chinese government published the *PRC Village Committee Organic Law (Trial)*, thus making the Village Committee the official system of villagers' self-government.[9] In 1983 Guangdong's Pearl River Delta adopted the government's policy of small-town-governing-villages. However, due to its special economic and geographical positions, most of the villages there had not formed Village Committees by 1987. Some villages only had a form of self-governed village management committees.[10] An example was Baixiu Village. In 1983, it formed the Baixiu Village government under the jurisdiction of Qingyang Region. In 1987, Qingyang Region was elevated to Qingyang Town, and Baixiu Village was elevated to Baixiu Administrative Region, but unlike villages in other provinces, it still did not have its own Village Committee.

Once the household agricultural responsibility system was implemented in the villages, the farmers regained control over rights to the use of the villages' resources and to production and trading, thus reducing the conflicts between farmers and cadres that had existed in the period of the people's communes. However, the economic reform and open-door policy were accompanied by a new kind of conflict. Although the Village Committees encouraged basic-level self-governance, the former people's commune cadres were still in charge of basic-level administration.[11] When they attempted to implement top-down central policies, such as collecting taxes, levying the agricultural tax in the form of grain, and birth control, they encountered enormous problems. In the era of the collectivized economy, these cadres had full control over the use, management, and trading of the villages' resources. The total earnings of the village were first calculated, and then taxes and expenditures for village facilities were deducted. The remainder would be divided among the villagers according to their work points. As a result, the resources for each farmer were determined by collective sharing. After the household agricultural responsibility system was introduced, resources were no longer dictated by cadres and villagers could control their own resources.[12] But the cadres still had to collect the taxes from the villagers. Whereas in the past, the taxes were first paid and then the remaining resources divided, now the taxes came out of the farmers' pockets. Currently, in some relatively economically backward villages, when the cadres demand taxes from the villagers, open conflicts often occur. Some cadres who are intent on meeting the demands of the central government may even use force to collect the taxes. This fuels the villagers' antagonism.[13]

In addition, under village self-government all primary construction work and development, like building roads and constructing irrigation systems, by and large, is financed and undertaken by the concerned villages. That is, the outlay for such projects is borne by the villagers. Often, the projects are very costly, and the farmers are largely dissatisfied, thus exacerbating their conflicts with the cadres.[14] In Sichuan, large-scale demonstrations required bringing in the army. In Guangdong, over 4,000 peasants tried to block a highway to protest the work of the village government.[15] As a result, the change in village governance brought about a new round of political crises.

In addition to the conflicts between cadres and villagers, there are also administration problems. Although the new rural economic reform and opening-up policy give peasants considerable leeway in administering their own affairs, in such things as farming methods, the sale and disposal of produce, and the right to use the land, the land is still collectively owned. Land allocations, still in the hands of the village cadres, are often a source of contention.[16] Some cadres take advantage of their power for personal gain. Many village cadres are thus nicknamed "local tyrants." They are sycophantic toward their superiors but behave tyrannically toward the ordinary people. Immediately after the economic reform and opening-up policy, such cadres made use of their positions to allocate land and make money through corrupt practices; they embezzled public funds from the sale of land, causing the villagers' pent-up anger to boil over. This has occurred in many places all over the country: both large and small demonstrations and protests have been staged in villages. Peasants have organized trips to higher-level governments, even to the central authorities, to lodge complaints against the local cadres.[17] In 1983 and 1984, villages near our fieldwork site staged protest marches, demonstrations, and road blockades. In Baixiu Village, the villagers discovered that the village cadres had dipped into the public coffers and had lined their pockets from corrupt dealings in landfill and construction projects involving publicly owned land. The villagers organized marches and visits to the central authorities.

Other problems arose because of the dubious qualifications of some of the cadres. Some village cadres had neither the managerial skills nor the level of knowledge to cope with their work. Others were too old, which affected their working ability. Many were ignorant of technological advances and concepts about the market economy. Not only did they not contribute to raising the level of the economy of their respective villages, they even paralyzed the economy. As a result, villagers grumbled about the cadres' inability to improve their standard of living, and this became another source of conflict. According to reports, anarchy paralyzed 20 percent of the village administrations in China. Some villages did not even have a Communist Youth League, Women's Federation, Security Committee, or similar basic-level organizations.[18]

Faced with instability in the rural areas, one of the leaders of the central government, Peng Zhen, put forward the idea of village elections. It is clear that the purpose was not to develop a political system of democracy in the villages; the goal of the government was merely to reduce tensions between village cadres and peasants through elections, with the aim of solving the problems of the economic and political governance of the villages.[19] For them, village elections were only a means, not an end. The debate over Peng's suggestion raged for more than three years. Finally, in 1987 the PRC Village Committee Organic Law (Trial) was passed, and some elections were held in 1988. But after the 1989 democracy movement, those leaders who had been opposed to the elections warned that they would promote instability in the villages. A new round of fierce debates ensued. As a result, the central government sent work teams to various villages to better understand the problems of management at the grassroots level. The teams concluded that the political situation at the village level was at a critical point. If the policy erred again, there would be further cataclysmic political crises. This greatly raised the temperature of the debate, so much so that the ongoing Village Committee elections were suspended. Some opponents of the elections argued that because the peasants were so dissatisfied with the central government, they might well pick cadres who supported the opposition. During this time, Communist Party elder Bo Yibo joined the debate and threw his weight behind the elections. His support had the effect of swaying opinions in favor, and in 1990 the central government decided to end the debate and to continue the elections for Village Committee.[20]

Implementation met with opposition not only from a faction in the central government but also from the local governments. Although the central government passed the Organic Law, six provincial governments did not pass the procedural stipulations for implementation.[21] Other provincial governments stipulated the procedural steps, but did not pursue elections vigorously. Resistance was even stronger at the county and town levels. Some county cadres declared that anyone implementing the Organic Law would be punished. In one survey, 60 percent of cadres at the town level in Hunan Province did not support the law. Other surveys reported that county and town cadres in the provinces of Shanxi, Shaanxi, Jiangxi, and Hebei in general did not support the law; some even prohibited it from being implemented. Some cadres pretended to support the law; others tried to fix the slate of candidates so that designated persons would be elected.[22]

There are a number of reasons why these county and town cadres opposed the Organic Law. First, in the past, village cadres generally had been picked by the county and town cadres themselves. They could be trusted to implement the policies of their superiors. This was thought to reduce management problems. But after the Organic Law was adopted, the elected officials were no longer under their control. In addition, Article III of the law stipulates that the role of the town government vis-à-vis the Village Committee is only one of guidance, no longer one of leadership.[23] Second, as pointed out earlier, after the economic reform and opening-up policy, the government often met opposition, sometimes leading to open conflicts, when it tried to collect taxes or other exactions. After the Organic Law was put into effect, those cadres elected by the people would try to protect the interests of the people and refuse to collect the various taxes due to the local governments. Even worse, it was feared that the villagers might use the law as a means of revenging former cadres for past wrongdoings.[24] Third, some local cadres regarded the peasants as selfish, feudalistic in outlook, superstitious, and lacking in modern knowledge, so it was believed that they could not possibly elect capable administrators.[25] Fourth, some local governments feared that local elections would boost the power of the lineages, unduly influencing village administration. This was one reason that the opposition to the law in the central government was so concerned.[26]

Although the Organic Law also met some opposition from the local governments, the peasants were very much in favor. There were reports that despite the local governments' reluctance to implement the law, villagers used all sorts of measures to pressure them to hold Village Committee elections, thus protecting their interests and ensuring a share of self-government.[27] Though the law had support at the grassroots level, the force of the opposition by local governments was strong enough to delay implementation. It was not until November 4, 1998, at the Fifth Meeting of the Standing Committee of the Ninth National People's Congress that the PRC Village Committee Organic Law was formally passed, eleven years after the passing of the trial law. The new law stipulated that by the year 2000, the one million or so villages in the entire country would have elections to choose members of

their Village Committees. Thus, a step was finally taken toward political reform at the basic level of the village.

Election of the Village Committee in Baixiu Village

In Guangdong, the Organic Law was popularized by selecting some trial
areas in each county to form demonstration units for other villages. Baixiu
Village was selected in 1998 as the test village for Qingyang Town, and
elections were held in December of the same year. First, in November, there
was an election for village representatives. The village is divided into six
groups (equivalent to the production teams from the people's commune era),
and these groups elected a total of sixty-nine representatives. Then, in early
December, the Baixiu Village Election Committee announced that election
of the Village Committee would take place in mid-December. The chairman,
deputy, and members of the committee would be directly nominated by the
village representatives.

Prior to the election, Baixiu Village, like other villages in Guangdong
Province, was not governed by a Village Committee. When the people's communes were abolished, Qingyang Commune was renamed Qingyang Town
and Baixiu Brigade was renamed Baixiu Village, thus beginning the town-
led-village administrative system. In 1987, after the central government enacted the Organic Law for trial implementation, it was hoped that Village
Committees would be formed all over the country and villages would administer their own affairs. Because of the special economic circumstances in
Guangdong, the law was not enforced in all villages. Instead, in some counties and townships, a new self-administered structure was formed—the administrative district. Baixiu Village Administrative District Committee was
structured similarly to other such districts under Qingyang Town: there were
seven members of the Baixiu Village Administrative District Committee responsible for the public and social affairs of the village. The committee
members were appointed, not elected, by the Qingyang Town government.
Moreover, these committee members were also members of the local Communist Party branch. The Party branch dealt with all village matters, including selling land, building houses, attracting foreign investment, constructing
roads, and implementing decisions of the central government. The Party
branch secretary (henceforth referred to as "secretary"), as in the people's
commune era, held both political and administrative powers. The Administrative District Committee was a self-administering structure in name only.
When the forthcoming election of the Village Committee was announced,
villagers had little idea about the real power of the new committee. But they
knew that through direct elections they could select persons whom they

thought suitable to run the village. Therefore, most of the villagers participated actively in the election. Villagers who held grudges against the secretary united to elect the chairman of the Administrative District, Wang Jianping,
as the chairman of the Village Committee.

Wang Jianping was about forty-seven years of age at the time of the
election. He belongs to one of the weaker branches of the Wangs in Baixiu
Village—the Tangzu branch. He is a villager in the third group. In 1978 he
was the leader of the second production team. In 1982, he became the general manager of an electric appliance factory. The factory was small, with
only a dozen or so workers. Capital investment was inadequate, the ma-
chinery rather primitive, and technical personnel in short supply. But by
1987, the number of workers had increased to more than two thousand.
Within a few years, the company was making profits exceeding US$4 mil-
lion. According to some villagers who worked in the factory, Wang's bril-
liant performance earned him a place on the Baixiu Village Administrative
District Committee in 1987. He was admitted into the Communist Party in
the same year. In 1990, he was elected deputy secretary of the Baixiu Village Communist Party branch and chairman of the Baixiu Administrative
District Committee.

When the election of village representatives took place, the majority of
villagers did not understand the roles of such representatives; nor did they
know that there would be an election for the Village Committee in December. Only after the date of the election and the methods to nominate candidates were announced did Wang Jianping and his supporters realize that the
secretary of the Party had managed to get elected a group of village representatives who then had control over the nomination process for the Village
Committee. The secretary himself would stand as the candidate for chairman. Wang and his supporters then attempted to understand the detailed regulations and procedures for the election, and Wang decided to run against the
secretary for chairman. In the process, they discovered that some of the regulations and decisions were unreasonable and unfair.

First of all, inclusion on the slate of nominated candidates was not the
only way to become a candidate. In the Guangdong Province Village Committee Election Regulations, it is stated that a person can be directly nominated as a candidate through an anonymous ballot by all eligible voters in the
village. This is known as *hai xuan* (literally, sea election).[28] Another way is to
have ten or more villagers nominate a candidate. The Guangdong Province
Village Committee Election Regulations stipulate that the election methods
be presented by the Election Committee to the villagers and then, after discussion, the method of election should be selected based on majority view.[29]
However, the Baixiu Village Election Committee never consulted with the

villagers before deciding that the candidates for the Village Committee should be nominated by the village representatives. Wang and his supporters wanted to make their views known to the Election Committee. To their surprise, they discovered that the members of the Election Committee, which had the power to supervise the election, had all been recommended by the secretary, and not by calling a meeting of the villagers. And the chairman of the Election Committee was none other than the secretary himself. The head of the Qingyang Town Village Basic Level Mediation Management Office was the wife of the cousin of the secretary; and the secretary himself was the deputy head of that office. Confronted by such interconnections, Wang and his group felt that it was useless to file a complaint, but they were not deterred. They continued to think of ways to expose the irregularities and unreasonable practices in the election.

One of the main tasks of the village representatives elected in November was to nominate candidates for the Village Committee. Wang and others found that more than half of the representatives did not receive more than half the votes, as required by the election rules. They also noticed that most of the elected representatives were supporters of the secretary. So they lodged a complaint with the Administrative District, pointing out that the election rules had been violated in the election of the village representatives. The secretary argued that calling another meeting of the villagers would be a waste of public money. Then Wang and the others went to the Qingyang Town government to argue their case. But the officials there asked them to let the matter rest. Failing to receive justice from the town government, Wang and the others could not but go one step higher, to the Guangdong Province Village Basic Level Mediation Management Office. On their way there, however, they were met by some members and staff of the Party branch who had been sent by the secretary to try to stop them, telling them not to blow the problem out of proportion, otherwise the reputation of Qingyang would be tarnished. Wang and his supporters ignored the threat, and proceeded to go to the Guangdong Office. When the Guangdong Office learned about the affair, it ordered that Qingyang Town postpone the election of the Village Committee by two weeks and once again hold an election for those representatives who had not received enough votes. Though a reelection was held for the thirty or so illegitimate seats, in general the same people were reelected due to the efforts by the secretary. Those elected were the main supporters of the secretary.

In addition to realizing that the election of some village representatives had been illegal, Wang and his supporters also learned that, at a meeting of the Town Committee to discuss the proper management of the grassroots level administrative structure, the secretary had announced that he would run for chairman of Baixiu Village Committee. Outraged, they aired their dissatisfaction with the Town Committee on several occasions, hoping that it would force the secretary to withdraw. Although the election rules do not prohibit members of the Election Committee to run for the Village Committee, the general practice is that if a member of the Election Committee decides to run for office, he or she will first resign from the committee.[30] Furthermore, it is government policy to draw a clear line between leadership of the Communist Party and leadership of all other public and social affairs. Generally, the government does not encourage chairpersons of village Party branches to run for the leadership of the Village Committee.[31] The secretary of Baixiu Village, noting the strong opposition by Wang and the others and mindful of his own problem of double identity, signed a statement withdrawing as a candidate for chairman of the Village Committee.

But the secretary insisted that if the villagers still voted to elect him as the chairman, he would accept the post. He knew that the fifth article of the Guangdong Province Village Committee Election Regulations states that villagers can vote for others not on the list of candidates, as long as those being voted for meet the requirement to be a voter.[32] It thus appears that the nomination of candidates is only a formality. Voters can cast ballots for whomever they wish, regardless of whether he or she is on the list. In fact, the *hai xuan* method was adopted during the second Village Committee election held in Baixiu: each voter had one ballot and could vote anonymously for whomever they wanted.

Although Wang was not nominated as a candidate from the village representatives, he knew full well that he could run for chairman of the Village Committee. He and his supporters launched a vigorous campaign. The secretary appeared to be out of the running, but his wife, in-laws, brothers, and cousins all proceeded to canvass votes for him. Interestingly, the two nominated candidates on the official list should have been campaigning for votes. But because they both served under the secretary and knew of his ambition and determination, instead they sent out pamphlets urging people to vote for the secretary.

During the early stage of the campaign, many villagers promised to support Wang Jianping. This was quite natural, for most villagers in Baixiu Village thought the secretary was using his power to line his own pockets. He had monopolized the economic development of the village. Most villagers hoped that through the election, they could pick some officials who were more upright and who would defend their interests.

Baixiu Village had implemented the household agricultural responsibility system since 1979. The approximately 5,000 *mu* of farmland were divided equally among villagers, meaning that each villager received about 2 *mu* of land. A family of four received about 10 *mu* for cultivation. Under this new system, the villagers regained the rights to the use of resources and to production and trading. Since Baixiu Village is located near Hong Kong and close to transportation routes, factories quickly sprang up there; by 1988 there were eight factories. It could be that the secretary understood the geographical advantage of the village. From 1988 on, he gradually reappropriated the right to make land contracts, the reason given being that this was necessary to develop the collective economy of the village. He filled in farmlands and built factories on top of them. Then he either rented or sold land to attract foreign investors to develop industry in the village. At the same time, he began all sorts of trading activities in the name of the collective. The secretary published the earnings from the first land sale, and the peasants who had formerly held the land-use rights were each given 100 yuan or more per *mu*. The secretary promised that the yearly rents from the factories, the management fees, and the earnings from trading in the name of the collective would be divided equally among the villagers. Most villagers supported this. In particular, the younger villagers saw this as a way to escape the grinding agricultural labor. Thereafter, most of the villagers in the six production teams hoped that their farmland would be selected by foreign investors, thus increasing their income and improving the quality of their lives.

The land sales peaked in 1991. Almost half of the farmland was sold to foreign investors or private developers. But the secretary no longer published the land sale prices, and he promised only that the villagers would receive the benefits from the sale. Also, the villagers were not allowed to take part in the decision making for the sale of the land. As they seemed to have lost their land-use rights and rights to production and trading, in early 1992, the villagers planned a protest march. The secretary and other cadres got wind of it and managed to hush things up. This incident was not unique to Baixiu. Between 1992 and 1993, many disgruntled peasants in the area who were unhappy that the cadres were monopolizing the rights to the use of land and to production and trading began to stage public protests.[33] The 1992 incident in Baixiu Village passed peacefully, and from 1991 to 1998, the buying and selling of farmland was still completely in the hands of the secretary; villagers took no part in decision making, nor did they know the value of the land sales.

But the villagers in Baixiu knew that for more than ten years, the secretary had been monopolizing the rights to the use of the land and its trading, and had used his position and power to amass huge amounts of money for himself. Two streets in the busiest industrial area in the village were nicknamed by the locals as "Secretary Street" and "Secretary's Lineage Street." The two streets comprised an area of about 20,000 square meters. This area was sold by the secretary in 1993 to an employee of the Qingyang government, at one-twentieth of its value. It was then bought by the secretary as a private individual in 1994 for exactly the same price. According to someone working in the Administrative District, the secretary saved 20 million yuan compared to the market value in this transaction. After buying the land, the secretary and his brothers and cousins started to develop it by building houses, factories, and shops for industrial and commercial use.

In addition the secretary's brothers formed an engineering planning and design consultancy. They cornered all contracts in the village involving engineering designs, quality assurance, and inspection of completed works and so on. Over the years the village spent a total of 4,000 million yuan for construction of factories, workers' dormitories, commercial shops, paved roads, and squares and stadiums. The consultancy company extracted 3 to 5 percent from each of these infrastructural projects for design, consultancy, and supervision services.

The villagers discovered that the secretary received substantial kickbacks from these projects. For instance, in 1993, the cost of construction per square meter to build houses or factories was about 300 yuan. The secretary signed a contract with a builder who bid 500 yuan per square meter. Thereafter, the builder complained to the villagers that the secretary was demanding a large kickback. In 1996, the secretary formed a construction company in the name of a few friends. The company received contracts for all the construction work, even though it charged more than other companies by about 100 yuan per square meter. Some villagers estimated that of the 600,000 square meters of construction work done in the village, the secretary received between 30 and 50 yuan per square meter in kickbacks.

In 1998 profits from industry and agriculture increased from 30 million yuan to over 40 million yuan. But each villager received only about 2,000 yuan. In the same year, the GDP of the village was about 200 million yuan, with net profits over 40 million yuan. The average income per capita was about 20,000 yuan. However, the secretary's economic report stated that because of the huge construction projects over the past years and with over 10 million yuan invested in infrastructure, the village actually incurred a debt of about 70 million yuan. At the end of the year, each villager received only 2,000 yuan from profits. It did not escape the villagers that in spite of this, the secretary and his relatives lived in mansions worth over a million yuan and they commuted in expensive imported cars.

Over the decade, the villagers in Baixiu were becoming increasingly angry about the way the secretary controlled village resources. In 1996 some villagers went to Beijing to lodge a complaint, accusing the secretary of corruption and working for his own self-interest. However, the trip was not successful.

Now the villagers saw the election as an opportunity to retake the right to manage their resources. During the early stages of the campaign, Wang Jianping was the favorite among most villagers.

When the secretary learned of Wang's increasing lead, he immediately launched a new round of campaigning. He met with influential elders and the household heads of the six groups separately and informed them of the benefits he would provide to their sons. He hoped that they would convince their family members to support the secretary. For example, most of the villagers in group one are surnamed Zhang. The three Zhang families constitute 90 percent of all the voters in that group. To garner their support, the secretary promised key members of these families certain benefits. He arranged for one member to become the general manager of a factory, and he promised that he would arrange for other family members to fill senior factory positions. Another Zhang family head had been the cashier for the Administrative District for five years and was due to retire. The secretary promised him that he could become the cashier of the Village Committee in exchange for his family's votes. In the second group, the secretary appointed the most influential elder of a Liang family to be manager of the Baixiu Village Sports Stadium. As a result of these campaign tactics, many villagers changed their minds and voted for the secretary out of self-interest.

Not only did the secretary provide benefits to some key persons, he also manipulated the election procedures. According to the regulations, if voters are working outside the village during the time of the election, they can vote by means of a letter of authorization. Initially, the procedure for getting a letter of authorization was very strict: the applicant had to present proof that he or she could not return to Baixiu Village on election day. According to the vetting procedure adopted by the Election Committee, only about 180 people qualified for such a letter of authorization. One day before the election, however, the secretary suddenly moved that any villager from Baixiu Village who was applying for a letter of authorization should be approved. On the day of the election, there were 440 such voter authorizations, among which over 300 of the voters voted for the secretary.

The six groups each formed an electoral district. Each district had supervisors and booths for the voters to mark their ballots in secret. The job of a supervisor was to ensure that the election be conducted fairly and there be no malpractice. On election day, the secretary increased the number of supervisors in each district from two to five or more. Some supervisors asked the villagers to show them their marked ballots to make sure they had voted for the secretary. Other supervisors filled out the ballot forms for the voters. One young villager in the first group said that he went into a private booth, intending to vote for Wang Jianping, but his ballot was forcibly taken away from him by his uncle and the secretary's name was marked instead.

When the results of the election were published, the secretary had received 950 votes and Wang 500 votes. The secretary thus became the chairman of the Village Committee. Although Wang and his supporters lost the election, they collected evidence of fraud and went to the provincial and Beijing authorities to protest the results, charging the secretary with rigging the election and using his official position to amass personal gain. Perhaps the secretary felt the heat from the public. At the end of 1999, he doubled the bonus from industrial and farming production distributed to the villagers, from the 1,000–2,000 yuan range to about 4,000 yuan each. In addition, approval of contracts was given over to the court in the township city where there was an open tender. The irony is that the popularly elected Village Committee met only once in more than a year. The affairs of the village, be they economic, public, or administrative, were all decided at meetings of the Party branch.

Family and the Election of the Village Committee

It appears that the network of human relationships in Baixiu Village still hinges on familial relationships. For this election, both the secretary and his opponent Wang Jianping canvassed the heads of families, *fang,* and lineage branches. If the heads agreed to vote for them, they did not have to approach the other members of the family, *fang,* or branch. They were confident that the other members would follow the vote of the family head.

As an example, when the secretary approached the Dezu branch, he first went to see the most influential family head of the four *fang.* Among the four *fang,* two had already declared their support for the secretary. But the heads of the other two still had a lot of complaints about the secretary. So the secretary settled the debts owed by one of the sons of one of the two *fang* and appointed him general manager of a medium-sized factory. He also gave the son of the head of the other *fang* a similar position. Consequently, the heads of these two *fang* promised to mobilize all their members and relatives to support the secretary. Wang approached the villagers through their families in much the same way, but he could not marshal the kind of material advantages or make promises similar to those of the secretary.

Because both campaigns were conducted along lineage lines, one result of the election was to cause a conflict between two of the branches of the Wang lineage in Baixiu Village—the Dianzu and the Dezu. Most of the members of the Hong subbranch of the Dianzu branch supported Wang Jianping; most of the members of the Dezu branch supported the secretary. Even before 1949, there had been in-fighting between the two branches for leadership of the village.

The nineteenth-generation Hong belonged to the Dianzu branch, that is, the branch claiming Dian as their ancestor. Hong and his father Lian had both been successful candidates in the national imperial civil service examinations during the reign of Qianlong during the Qing dynasty (around 1740 A.D.). That made the branch the most prominent among the lineage. In the case of the Dezu branch, the genealogical record shows that for some reason, De's father's family met with lean times and did not prosper. De was the third child. He did not like studying and he preferred the martial arts. The descendants of De claimed that De fell in love with Hong's maidservant and she returned his love. But Hong accused De of harassing his maidservant. Because of the prestigious position of Hong and his father, De was eventually brought to the county court and tried. Though he was not convicted, De felt that the influence of Hong and his father was such that he could no longer stay in the village. He eventually found his way to west Guangdong and entered Yue Military Academy. He later took part in the imperial examinations and was successful as a candidate in military subjects. With his newfound power, De moved back to Baixiu Village, and the status of his branch, the Mozu, was raised. But the past feuds still haunted them, and the Dianzu and the Mozu were still enemies. De's four children were all successful in the provincial civil service examinations, whereas Hong's six children were not. Even today, De's descendants say that Hong fathered six worms and De four dragons. De's descendants built a temple called The Venerable De's Temple and set themselves up as a new line of progeny, the Dezu branch. Because Hong's descendants did not build a new temple to honor Hong, his progeny still belong to the Dianzu branch. In the past, the two branches fought one another, vying to become the main force in the village. After 1949, both branches were declared landlords or local tyrants, and because of their class status, none of their members was ever appointed as a village cadre.

Members of the Dezu families supported the secretary and members of the Dianzu families supported Wang Jianping. There were many conflicts between them during the campaign. In one instance, one of Wang's supporters, Wang Weijie, who belongs to a Dianzu family, led a very visible campaign to become the group leader of the fifth group, which consists largely of descendants of Dezu. But the Dezu regarded this move as a challenge, which fanned their animosity. The contest between the secretary and Wang became a contest between the two groups of descendants. They openly heaped insults on one another in the marketplace, and plastered street walls with big-character posters attacking one another.

After the election, I visited Uncle Cheng, a Dezu, to gain further under-standing of the election. Uncle Cheng is one of the most influential Dezu elders. He was also on very good terms with the secretary. Before the elec-tion, his son, generally considered a good-for-nothing, was appointed the general manager of a factory. Maybe that is why Uncle Cheng campaigned so hard for the secretary.

After a few pleasantries, I asked him about the election. He responded:

"In this election, they [the Dianzu] were routed. They're useless. How can they expect to get the better of us?" Uncle Cheng might see many faults in the Dianzu people, but he seldom openly expressed his opinion in such a way. This time he could not suppress the joy in his heart.

"They [the Dianzu] are now hiding in their homes. They dare not come out and dare not go to the Administrative District [meaning the office of the new Village Committee]. They have lost face! One week ago, the Fraternity Association held a function. People had to go to the Administrative District to get permission to use a bus. None of those Dianzu dared to go see the secretary. They knew the secretary would not give them face. In the end I had to go. What a load of crap about defeating us; now see who's defeated!"

The chairman of the Baixiu Fraternity Association is Uncle Xing, the most respected elder of the Dianzu branch. When the society was first formed, Uncle Cheng and Uncle Xing both ran for the post of chairman. It was Xing who won and he had been chairman for two consecutive terms. Because during these terms Uncle Cheng had only served as a secretary, he always harbored a grudge against Xing.

"They [the Dianzu] aren't united," said Uncle Cheng. "Getting a has-been trickster to return is useless. He lost to the secretary before. What new tricks could he have?" The "trickster" he was referring to was Wang Huihuai, who had run against the present incumbent for the job of secretary in 1974 and lost. Some villagers reported that he felt utterly dejected after that, but after a few years, when the economic reforms began, he returned to Qingyang Town to do business.

"Did he campaign for Wang Jianping?" I asked.

"He didn't come back just to get votes for Wang. He is a Dianzu and he worked for the Dianzu families. Most of them supported Wang. If Wang wins, that means the Dianzu win. That boy Jie, see. He ran for group leader and did he not lose to a Dezu? They were really bitter, writing big-character posters to attack us Dezu. They branded us as 'rat feces'!" As Uncle Cheng spoke, he became more and more worked up. I learned that Uncle Cheng had not passively accepted the insults. He too put up many big-character posters in the marketplace that attacked the Dianzu.

"Do you not think they might have a point?" I pressed. "They think that since more than 75 percent of the village is made up of Wangs, it stands to reason that a Wang should be elected secretary of the Village Committee. But you threw your weight behind the secretary." The family of Secretary Zhou Weiquan had lived in Baixiu for only three generations.

"The chairman of the Village Committee should be someone with ability, not necessarily someone surnamed Wang. We should elect someone capable. If you elect your own kind, how can you run the village well? As for Wang Jianping, he has no ability, and he has no experience in public office, yet he still wants to be chairman. How can that be? The secretary had already advised him to be patient, to run for chairman the next time around, but no, he wanted it right away. He even boasted that for sure he would defeat the secretary."

It seemed that Uncle Cheng had changed quite a bit. Several years earlier, I had accompanied him to Bailang Village to attend a meeting of the Wang lineage. I can clearly remember how Uncle Cheng and other elders then admired the Wang lineage, because the secretary and the cadres there were all Wangs. At the time he had said that having your own relatives as leaders was much better than having outsiders. One's own kind will look after you better. On that occasion, they grumbled that their own secretary was not a member of their lineage and he ignored the interests of the Wangs in the sale of the land.

"They only made up the excuse of electing your own kind because they wanted to get Wang Jianping elected. An election should be a rational process. Why talk about insiders and outsiders? Isn't this being divisive? The main reason why they lost was that they are not united. We are different: we're united, of one heart, and we're not split over personal interests."

"But aren't there Dezu people who didn't support the secretary?" I knew that there were two Dezu descendants, Ah Sheng and Ah Zhao, who openly objected to the Dezu people rallying behind the secretary.

"Those two are the 'running dogs' in our Dezu line. They didn't help their own kind; instead they worked for others. Well, it can't be helped! It's said that they owed people [meaning Wang Jianping] over 100,000 yuan! So they were led by the nose. There's that fellow Qiu Ji too [Wang Zhenqiu, Ji being a familiar personal suffix]; he was with those two [Ah Sheng and Ah Zhou]. One evening, we held a meeting in the Dezu ancestral temple to talk about the election. The secretary was also present. That fellow Ah Qiu, he knew the meeting was about the election, yet he asked Ah Sheng and Ah Zhou to join in. He clearly wanted them to attend as moles. I think Ah Qiu is not reliable."

From what I could gather from asking around, Ah Sheng and Ah Zhou were not well acquainted with Wang Jianping, nor had they borrowed money from him.

During the interview, Uncle Cheng seemed to be very content. The reason might not only be that the person he and the Dianzus had supported was elected; more probably it was because they felt that they had dealt a blow to the Dianzu people in the election. Thus, he now felt relatively more important. Any activities organized by the Wang lineage or the Fraternity Association had to go through him in order to receive the secretary's blessing.

Wang Zhenqiu, who was lumped together with Ah Sheng and Ah Zhou in the eyes of Uncle Cheng, took quite a different view of the election. During our first interview after the election, he called out to me loudly as soon as he saw me, "You know what? My surname has been changed! My surname has been changed! I am no longer a Wang. You know what I mean?"

I was taken aback by such a remark, but after a little while understood his point. "My surname is now Zhou; it can also be Cao. At any rate, it is no longer Wang!" The secretary is surnamed Zhou. The villagers had compared him to Cao Cao, a historical figure during the Three Kingdoms period noted for being crafty and sly; they called him by that name behind his back.

Uncle Qiu did not support Wang Jianping, but like some old villagers, he hoped that a Wang would become the chairman of the Village Committee. He was quite clear about the secretary's character, so he did not want to support him. He regretted that his Dezu relatives backed him. Although he tried to remain aloof during the election and did not join the activities organized by the Dezu to garner votes for the secretary, he was still ridiculed by other Wangs, who criticized him for being so concerned about Wang affairs yet not supporting a Wang. During the election, people from the Dianzu branch often taunted him, calling him "Zhou Zhenqiu," implying that he had forgotten his ancestors. This hit him hard and as a result he often stayed at home alone when the others were campaigning. He hoped thereby to avoid the controversies. But still he could not. Because he did not actively canvass for votes with the Dezu and because he informed Ah Sheng and Ah Zhou about the meeting in the Dezu ancestral temple to support the secretary, the others suspected him of having joined Ah Sheng and Ah Zhou's side.

"Everything is lost! To support someone with a different surname—that's the same as changing one's own, isn't it?" Uncle Qiu said petulantly.

"That means you supported Wang Jianping?" I asked.

"How could I support him? I am a Dezu descendant. Moreover, I don't know if Wang is all that reliable. Don't people say 'It is better to live with a well-fed tiger than a hungry wolf'? How could I support Wang Jianping?" Some people surmised that the secretary was a "well-fed tiger" who had his fill of gains, whereas Wang Jianping was a "hungry wolf" who had nothing.

If Wang were to be elected, he might be even worse than the secretary in terms of filling up his own coffers.

"That's why you voted for the secretary?" I asked.

"There was no other way. We were family. If I did not do so, I would be branded a rebel of the Dezu. During the election, the Wangs of Baixiu Village were split into two camps. They are no longer united." Uncle Qiu was obviously saddened. "I think the matter will not end like this. They [Wang Jianping and his followers] won't just lay down their weapons. They will fight on, because the truth of the matter is that the secretary is corrupt!" He stopped, one hand holding a cigarette, the other making a movement as if to accept bribe money.

Uncle Qiu was not on the side of the secretary, and he dearly hoped to have a Wang as chairman, yet because he is a Dezu, he thought that he should follow the decision of his branch. Such a view was not confined only to the older villagers. Even some of the young villagers thought they should support the candidate favored by their branch, and stand together with their family members by going along with the collective decision.

Ah Chuang, about thirty years old, is in Uncle Qiu's group. When the group elected its leader, Ah Chuang wanted to vote for a certain candidate called Ah Lian who was quite fair and capable in getting things done. But the other candidate was a member of the same family group, so he had to vote for him.

As Ah Chuang put it, "When you elect a candidate, you must elect someone belonging to your *jia tou* [family group, with the family relationship traceable to five generations]. That's just the way it is. Even if your own person is not capable, you have to elect him. He is family!"

"Did the man from your *jia tou* win?" I asked.

"Of course. We have more than forty people in the group. On top of that, there are relatives of our *jia tou*. He won by a wide margin," Ah Chuang answered.

"Since Ah Lian is stronger in many ways than the candidate from your *jia tou*, why couldn't he win?"

"There are only a few people from Ah Lian's family in our group, and he could not get votes from others, so he was defeated. An election is not about ability or performance. It is about family relations." Ah Chuang further pointed out that if a candidate is not a member of one's family, but the family decides to support him, then all family members must lend their support. Whoever does give support will be ostracized. Ah Chuang's view is quite representative of Baixiu villagers.

But there were some villagers who stuck to their personal views and did not follow the decision of their families. Ah Sheng and Ah Zhou are two of the

Dezu descendants who publicly declared they would not follow the decision of their branch.

Ah Sheng and Ah Zhou, branded "running dogs" by Uncle Cheng and other elders, are descended from, from the family of the third brother of the Dezu branch. They are both over thirty years old and married. Ah Sheng is a motorcycle driver, who mainly transports villagers in and out of Baixiu Village. He makes a reasonably comfortable living. Ah Zhou sells frozen meat in the next village and is quite well off too. In an interview, the two of them expressed their views about the election.

"Over the years, people have come to know what kind of a person the secretary is. Why must we support such a fellow? Of course we didn't support him, but we did not betray the Dezu. Why did they call us rebels of Dezu and even 'running dogs'? We are direct nephews of Uncle Cheng, from the same grandfather; why use such vicious terms?" Ah Sheng became more and more heated as he spoke. "The day Uncle Qiu told us that the Dezu would hold a big meeting—as you know, Uncle Qiu never speaks clearly. If he had told us clearly that it was to be a rally for the secretary, we would not have gone! All that nonsense about us being moles!" I had heard that because of the election, Ah Sheng and Ah Zhou both had had many public heated rows with Uncle Cheng in the marketplace.

"We are Dezu people. That does not mean we follow every decision of the Dezu elders; we only support what is reasonable. We also supported Ah Ming [Wang Qiming] as our leader [of the fifth group]. You think Uncle Cheng and the others are right? If they had not gotten some personal advantages, why were they so active on his behalf? Several years ago, when we rebuilt the Dezu ancestral temple, didn't Ah Ming get into trouble with the other *fang*?" Ah Sheng continued.

"Shi Quan [Wang Qiming's father] and Ah Jin both got something out of it. That's why they asked all their family members to support the secretary," Ah Zhou added in agreement.

In 1995, when the descendants of the four families of the Dezu discussed rebuilding the Dezu ancestral temple, Uncle Cheng was asked to sell his land-use rights to a quarter of the land on which the old temple was located, for 20,000 yuan, a discount on the then current land price. It would be regarded partly as his contribution to the familial cause. But Uncle Cheng objected. He wanted a piece of land in exchange for his share of use of the temple land; only then would he consider it a fair deal. For this, he ran afoul of the other families. This internecine squabble delayed the construction of the temple for more than a year. It was reported that finally a member of the third family, in order to hasten the rebuilding of the temple, exchanged a piece of residential land he had bought a few years earlier for Uncle Cheng's land rights, thus ending the stalemate. That piece of residential land was worth about 80,000 yuan. Most people in the branch were upset with Uncle Cheng's selfishness. On the day of the feast to celebrate the inauguration of the new temple, they distributed a list of donors to the rebuilding of the temple and a statement of accounts. In both, Uncle Cheng's insistence on exchanging his right to the use of the temple land for a piece of land was recorded, a move aimed at making him lose face.

"They went over to the secretary's side for their own selfish gain. They knew the secretary was greedy for money, and they described Wang Jianping as a 'hungry wolf.' It was their own business if they favored the secretary, why force other people to vote in the same way? If you didn't do what they said, they called you names. It's really ridiculous!" Ah Sheng huffed.

"As a matter of fact, I believe that not all Dezu people supported the secretary, but dared not admit it openly. If Uncle Cheng thought that every Dezu member was on his side, why did he ask Ah Ming [Wang Qiming], Ah Hong [Wang Weihong], and Ah Chao [Wang Yongchao], and others to stand outside the private booths to inspect the Dezu people's votes to make sure they had voted for the secretary?"

Ah Sheng and Ah Zhou made it quite clear that they would not vote for the secretary, yet they did not attend the caucus for Wang Jianping's supporters. Ah Sheng never canvassed votes for Wang Jianping and Ah Zhou only occasionally distributed leaflets for Wang Jianping when he had some free time.

Unlike Ah Sheng and Ah Zhou, a sizable group of villagers did not publicly oppose their elders or family heads. Still, they did not want to follow the instructions of their elders. They still voted for their preferred candidate. Unfortunately for them, however, at the polls their ballots were checked by their elders or relatives. For example, at the polling station for group one, nearly a majority of the ballots were taken away by the elders to change them to votes on behalf of the secretary. Only a few of the voters insisted on filling out the ballots by themselves. Some young villagers complained that they were not allowed the right to vote for the person of their choice. Among this younger group, Wang Guizhen came up with her own way of making sure her ballot did not fall into the hands of the secretary's supporters.

One morning after the election, I visited Wang Guizhen. Her father-in-law, mother-in-law, and husband all happened to be at home. During our interview, I asked them whom they had supported in the election. Her father-in-law and husband both answered that they had voted for the secretary. Ah Zhen did not reveal for whom she had voted. But she did say: "Of course, whoever has money and power will be elected. An election in a rural village is always like that."

During our meeting, Ah Zhen did not want to say much about the election, probably because of the presence of the others. So I invited her to Li Zhichao's house to talk.

At Li's house, Ah Zhen talked to her heart's content. She had supported Wang Jianping. On election day, she went to the polling booth of the fourth group. First, she avoided those people who wanted to fill out the ballot form for her. Then, she went into the private booth to vote, and immediately proceeded to place her ballot in the ballot box.

"At the polling station, there were a lot of people who worked for 'Cao Quan.' And those who were supposed to supervise the station were not there to stop any wrongdoing; they were there to make sure their family members had filled in the name of 'Cao Quan.' When I was about to cast my ballot, a supervisor came over and wanted to see whom I had voted for. Trying to avoid trouble, I said to him, 'Don't be so nosey!' Then I quickly left him and put the ballot into the box." The way Ah Zhen told me this story showed that she felt herself to be quite astute. "If they knew I had voted for Wang Jianping, they would not have let me go until I changed my vote. I didn't want to get into a quarrel with them. After all, they're members of my production team. Some of them are even relatives of my husband."

Ah Zhen went on talking about the election day events. "On that day, many people were under pressure to vote for the secretary. Despite all their tricks, he still received only 400 or so more votes than Wang Jianping. Without such tricks, Wang certainly would have won. When the secretary announced the end of the election day, he seemed unable to concentrate, and said something like 'Thank you for coming. Now you can all go home and eat your own meal'! What a vulgar thing to say on such an occasion!" Ah Zhen was becoming more and more worked up. "And he really meant it! He was asking his supporters to go home and have their dinner. Serves them right! They were all expecting the secretary would host them to a banquet at Qingyang Hotel after the election."

Ah Zhen's husband's family had all supported the secretary, and her father-in-law spared no efforts in getting votes for him. Therefore, during our first meeting, in the presence of her in-laws, Ah Zhen did not say much about her own views. "Since father-in-law was on the side of 'Cao Quan,' of course he told us to vote for him. We exchanged some cross words, but I found that to be useless. So I put on the appearance of agreeing with him. In that way I did not offend him."

"Did you have any arguments with your husband?" I asked.

"He had just returned from another town. He had not yet completed his resident's registration form, so he was ineligible to vote. Anyway, he is the sort of person with no strong personal opinions. Even if he asked me to side with the secretary, I wouldn't have listened to him. He has his views, I have mine. Why copy his?" Ah Zhen answered in an upbeat tone.

"Does your father's family support Wang Jianping?" I asked.

"I'm not sure, and I didn't ask them. Sister seems to have voted for Wang. We [her father's family] won't vote for 'Cao Quan.' He never helped any one of us. Even if he had, it was our money. Wang Jianping is a much better person than 'Cao Quan.' He is capable and he is fair in his dealings, unlike 'Cao Quan' who has been monopolizing the village's economy as 'one dragon.'" (The expression signifies a continuous line of control.) Ah Zhen meant that the secretary gained personally through the sale of land and construction contracts and that his relatives all had a hand in the businesses of the village. That is why she thought that the secretary's wealth should belong to the village.

There are many villagers who, like Ah Zhen, know the kind of person the secretary is, yet they did not support Wang Jianping. In the past, Uncle Quan and Uncle Zhao, descendants of the second son in the Dezu line, often criticized the secretary for lining his own pockets through his position. Still, they voted for him and actively garnered votes for him. It was said that they changed their attitudes toward him because the secretary did favors for their sons. Uncle Zhao's son Wang Weihong had worked elsewhere for a number of years but he could not find a good job. Eventually, prior to the election, it was arranged that he become the general manager of a rather large factory in Baixiu Village. Uncle Quan's son Wang Qiming at one time had been a general manager of a foreign-owned factory. More than half a year before the election, it was found that he had embezzled about 400,000 yuan in company funds. With the assistance of the secretary, he was not prosecuted.

Both Wang Qiming and Wang Weihong vigorously helped the secretary to gather votes during the lead-up to the voting. On election day, they were appointed supervisors at a voting station. But they did more than just supervise: they stood at the entrance to the polling station and persuaded villagers of the fifth group to vote for the secretary. They even went into the booths and asked members of the Dezu family to write down the secretary's name on their ballot form. Also, they checked the completed forms to make sure the voters had voted for the secretary as promised. After the election, Qiming was appointed the general manager of a large-scale foreign-invested factory, with a substantial salary and benefits. In 2000, his father became one of the managers of a newly built market, about 16,000 square meters in size, in Baixiu Village. Also in 2000, Qiming spent about 200,000 yuan to establish a lucrative restaurant near the sports stadium.

In an interview, Wang Qiming and Wang Weihong told me their views about the election. "An election is about advantages. If you give me advantages, I will support you. An election is a practical thing. It is not about what you talk about in Hong Kong, like democracy."

"Is that why you did not vote for Wang Jianping?" I asked.

"His power base is far thinner than the secretary's. If he were elected chairman of the Village Committee, I am sure that our economy would go downhill. He has no personal network either in the town or in the township. When he wants something done, how can he be expected to get it done efficiently? The secretary is different. Everywhere he goes, people will give him face. People are pragmatic creatures. If the villagers thought about their own interests, why would they elect Wang?" Then he added, "See, once the secretary was elected, our yearly bonus doubled; we got over 4,000 yuan."

"But didn't Wang's platform promise to increase the yearly bonus to 6,000 yuan?" I queried.

"That's impossible. It is not realistic. How could people believe that?" Ah Hong further explained, "You see, if we Dezu people were not so united, why would the secretary come to ask for our support? Conversely, if the secretary were not an able person, we [Dezu] would not support him."

One can see that Qiming and Ah Hong's views are quite different from those of Ah Zhen and Ah Sheng. The reason could be that in the election Qiming and Ah Hong got many advantages, so they did not touch on the question of the monopoly of the village's resources, or other unfair practices. For them, elections are purely a business transaction.

In the election Li Zhichao behaved very differently from Qiming and Ah Hong. He refused the advantages offered to him by the secretary, and together with his second brother, threw his weight behind Wang Jianping.

When the election was first announced, Zhichao had an open mind about with whom he would side. But he had a vague idea that supporting the secretary would bring him benefits. He told me how some motorcycle drivers had made use of the opportunity offered by the election to go to the Administrative District to stage a protest.

"This year, a group of motorcycle drivers from other provinces came to the village and started taking business away from us. We were upset, of course, so we went to see the head of the Security Committee, hoping he would do something to drive away these intruders. One of these drivers turned out to be a relative of a sister-in-law of the head of the committee. So the committee refused to listen to us. After quarreling with him, we called all the motorcycle drivers to drive to the committee to protest. Seeing the approaching motorcade of protestors, the head of the committee went into hiding. But we were not deterred and drove instead to the Administrative District office to see the secretary. When he learned about our complaint, he promised that he would not allow motorcycle drivers from other provinces to work in the village." Zhichao said that since this incident took place during the election period and the secretary wanted to win votes, he quickly solved the problem. He could see there would be further advantages as well if they supported the secretary's bid for election.

Two weeks before the election, I found that Zhichao had undergone a change in attitude. He now sided with Wang Jianping. After the election, I went to talk to him to find out the reason for his change of mind.

Zhichao was frank about his lack of a definite choice at the beginning of the election. He sensed that the secretary was so rich and powerful that he would certainly win. Zhichao's second brother, Zhiquan, worked in the same factory as Wang Jianping; they had worked together for a number of years and had become close friends. During the election, Zhiquan had mobilized his family members to vote for Wang; furthermore, he went around the village soliciting votes on Wang's behalf. One day, a trusted subordinate of the secretary, who was a cousin of Zhichao, tried to bribe Zhichao to convince his second brother and the other Lis to give up backing Wang Jianping.

Zhichao passed the message on. "When my two brothers came to see me, I told them that if they supported the secretary, those of us who were not factory managers could become managers; those of us not in the Communist Party could be admitted earlier. But if we voted against the secretary, even a manager could be dismissed." He made an effort to persuade his brother not to support Wang Jianping. His second brother did not listen to him; instead he tried to convince Zhichao to vote for Wang.

"You mean to say you supported Wang mainly because of your brothers?" I asked. "Well, they're my blood brothers. If they supported Wang Jianping, there's little I could do. After a little while, my cousin came to see me again. He told me I could have a taxicab and become the manager of a factory. He hoped that I would accept this offer and convince my brother and the other Lis to vote for the secretary. Finally, I rejected his offer." This reminded me of an incident during the time of the election. Zhichao had telephoned me in Hong Kong to ask about the cost of studying abroad. I was quite puzzled by the call. Now I think I understand. That call was probably made after his cousin's second visit, when he promised him more advantages if Zhichao were to support the secretary. Zhichao might have been thinking about his son's future and was inquiring about study abroad when he was still in a state of indecision.

"When my cousin made those proposals, I did not accept them. Still I thought about them. I was afraid that if the secretary won, he would take revenge. I decided that I would ask my third brother not to take part in Wang Jianping's campaign." Zhichao thought that was a pretty smart compromise. He told me that other people had the impression that his third brother was not in Wang's camp.

"Even Ah Cheng [Uncle Cheng] did not know the truth. He said to me that my third brother was sensible about not joining Wang's gang. . . . If we lost, my third brother would not be implicated. So there would be someone in our Li family who could carry on." Zhichao told me that during the election period his second brother felt some pressure in his factory. People asked him to mind only his own business. His third brother was not harassed in this way.

As things turned out, Wang Jianping lost the election. But he and his supporters could not accept the results. They collected a lot of materials about the way in which the secretary controlled the election, about his corrupt and deceptive practices, and about the way he used his position for personal gain. Then they wrote an exposé to complain. They went to Beijing intending to sue the secretary. At first they thought that the only way to approach an official of the central government was through a personal network, so they fell easy prey to some crooks who swindled them on the pretense that they could put them in contact with the highest officials in charge of prosecuting corruption; the crooks made off with more than 700,000 yuan of their money. The presence of such swindlers is not surprising. Because many peasants go to Beijing to sue local officials, the crooks can have a good time. In spite of being deceived, Wang Jianping and the others did not give up. For more than half a year they persisted, pursuing all sorts of channels and lodging complaints against the secretary with over ten departments at the provincial and central government levels. In the end, nothing came of it. When Wang Jianping and his followers realized they were not making any headway, they were crestfallen.

Around the end of 1999, I returned to Baixiu Village. It had been almost a year since the election. Memories of the election and the attempts to sue the secretary were fading from the minds of the villagers. Wang Jianping and his supporters went about their own business and had not met as a group for some time. Zhichao continued to drive his motorcycle. When he thought back to the election, however, he felt regretful.

"We put in so much effort, and the result is the same as if we had not done anything! I was stupid. I didn't think clearly enough. The secretary is rich and powerful, and the result was predictable. I should have accepted his offer of a taxicab. Now it would be worth more than 300,000 yuan. That's a lot of money! How long and how hard do I have to work to save that amount?" he said, with a heavy sigh.

"I used to laugh at those people who helped 'Cao Quan.' I had said that 'Cao Quan' would not keep his word. But look at Ah Wa in the sixth group. Before the election, he was heavily in debt and he had to ask me to lend him some money. Shortly after the election, he had a pretty mansion built in the resort area. And those other people! One became the manager of a square, another the manager of a market, and still another the manager of a factory, and so on. Anyone who worked for the secretary's campaign reaped great profits. I am the stupidest person around here!" He was especially regretful when he thought about his son, who, if he went to university, would be a huge financial burden on the family.

One day at the beginning of 2000, I had dinner with Zhichao, Qiming, and Ah Hong. After walking me back to the hotel, Zhichao told me more about what had happened during the year after the election. The topic touched on Qiming and Ah Hong, both of whom had become general managers of factories with good salaries and bonuses. Zhichao could not help but blame himself again. "That was the biggest mistake I've made in my whole life. The biggest mistake! I won't be able to earn that much in my entire life." He inhaled deeply from his cigarette, lying on the bed and sighing.

Conclusion

During this election, the villagers of Baixiu were unable to exercise the political rights given to them by the government to put the tenets of democracy and autonomy into practice. This was not an isolated incident. It is quite common among rural societies that have had only ten years of experience in basic-level political reform. Some scholars think that this is because villagers do not understand the broader concepts of civic responsibilities and rights or individual political rights. So they cannot make rational and autonomous decisions.[34] Other scholars point out that a more thorough political reform is required to make peasants into rational and autonomous citizens and to achieve true political autonomy.[35] Of course, no one will deny that external political factors will have important effects on the election of the Village Committee. It is argued here that the concomitant cultural forces exert a certain countering influence, such that the villagers encounter obstacles in realizing their rights and autonomy. In the following paragraphs, we shall use the concepts of "insiders" and "outsiders" in lineage culture, material interests, and individual autonomy in an analysis of the election of the Village Committee, to determine why democracy was not achieved and to evaluate the development of individual autonomy among the Chinese.

Within the moral values of familial culture, human relations are polarized. In the familial network, members of the family group are "our own people" or "insiders." Such people can be trusted, help one another, and work cooperatively for the common good of the entire lineage. People outside the kinship network are "outsiders" who are motivated by selfishness. They make use of one another.[36] In actual life, though, some people with traceable familial relationships are regarded as "outsiders." A lineage is not a unit with a well-defined boundary; kinship relationships are relative, and the members' roles are relative as well.[37] Members of a lineage often break up because of arguments over the sharing of inherited ancestral assets.[38] Take the Wangs of Baixiu Village as an example. The Wang family has divided itself into some twenty branches. These Wangs all live in Baixiu Village and all have the same surname, and yet they do not always regard each other as "insiders" because of the different lineage branches. Some branches clashed with others in the past due to conflicting interests, or due to violent criticisms and torture during the Cultural Revolution. These branches are hardly "insiders" with one another; rather, they are antagonistic. The feud between the Dezu and Dianzu is a prime example.

Even within the same branch, not all members are regarded as "insiders." Take the Lingzu (ancestor Ling's descendants). There are now more than ten generations of Wangs, offspring of ancestor Ling. In other words, some members have to trace their ancestors back ten generations before they find they came from the same ancestor. Because of differences in closeness in the familial relationship, these Wangs do not necessarily see some members as "insiders." After 1949, the ancestral halls that held the branches together were mostly destroyed and the ancestral properties were confiscated by the government. The members of the same branch were no longer closely knit. Today, Baixiu villagers regard the Wangs within five generations of familial relations as belonging to their own "family group" in their daily lives. Only these people are considered "insiders." According to the lineage culture, when "insiders" meet any difficulties, they should unite to help one another. When there is a conflict of interest, there should be trust. Other people with the same surname and belonging to another "family group" will be treated as "outsiders."

In this bipolar division, it is difficult to foster a cooperative contractual human relationship that works toward common interests.[39] During the election, one can find two effects of the familial culture. First, the bipolar human relationship in familism obstructed the villagers from making a rational choice to express their individual political will in the election for the head of the Village Committee. As a result, the election became a business transaction. Second, by exercising their individual political rights through the election process, the villagers had an opportunity to reclaim their right to use the village resources on the one hand, and to manage the public affairs of the village more rationally and distribute the resources more fairly on the other. Unfortunately, the moral system of the familial culture took precedence over the concept of justice in public affairs, and to a large extent directed the villagers' choice. As a consequence, individual autonomy did not develop properly during this election.

We will first try to show how the bipolar human relationship in the familial culture made the election of the Village Committee into a business transaction. Once the election was announced, many villagers hoped to seize the opportunity to exercise their right under the election laws to manage the village themselves—to ensure an equitable use and distribution of resources and to manage public affairs properly. During the election, when some villagers realized that the secretary was manipulating the election procedures, they immediately went to the Qingyang Town government to make a report. They even went to the Guangdong Province Village Basic Level Mediation Management Office to complain that the village representatives were not being elected according to regulations. As they were able to cite the relevant regulations in the Village Committee election rules, some village representative elections were annulled and a new election was called. Such complaints were not confined to Baixiu Village; in villages all over the provinces of Guangdong, Hunan, Shanxi, and Hebei similar complaints were lodged.[40] Villagers used the rights given them by the government to complain to county or provincial governments. They wanted the election to proceed according to the principles implicit in the rules. In fact, such complaints did exert pressure on county, provincial, and central governments to consider the legitimacy and fairness of the election procedures. This way of applying to a higher authority to complain about the election procedures is, in the view of some scholars, the result of changes brought about mainly by the economic reforms over the past twenty years. The market economy has brought with it a villagers' awareness of such notions as rights, fairness, and justice. They have achieved a greater understanding of laws and regulations.[41] However, these concepts did not seem to have been realized in the context of the election of the head of the Village Committee.

When canvassing for votes at the early stage of the campaign among his family members and kinsmen, Wang Jianping had the support of many villagers coming from different *fang* and branches of the Wang lineage. But not every one of them regarded Wang Jianping as an "insider." Those who saw him as an "outsider" did not believe that once elected chairman of the Village Committee, he would govern the village justly and fairly. Even though they verbally supported Wang, in their hearts, they thought that the candidates and the electors were making use of each other. Wang Zhenqiu was one such person with these views. He had many reservations about the character of the secretary, but since Wang Jianping did not belong to the Dezu branch, he suspected that he was running in the election for selfish motives. The phrase "well-fed tiger and hungry wolf" circulated during the campaign to describe the differences between the secretary and Wang. People thought that once they ousted the "well-fed tiger" who had filled his coffers, then the "hungry wolf" would take its place, and since it was hungry, it would grab at anything it could lay its hands on, and the corruption would be even worse.

This distrust of Wang helped the secretary, who threatened and bribed the villagers during the second round of vote-getting. As a result, many villagers changed their minds and supported him. It is an undeniable fact that the secretary's wealth and power led some villagers to abandon their desire to regain the right to administer Baixiu Village. The traditional familial culture was another factor contributing to the villagers' change of attitude.

Though most villagers in Baixiu are surnamed Wang, they lack a relationship of trust among themselves. They are suspicious of one anothers' motives. The building of trust between "insiders" and "outsiders" has to be premised on concepts like honesty and a social contract. Without them, cooperation will be fragile. In the twenty or so years of economic development, there have been opportunities for the development of a contractual relationship among the people to cooperate to achieve common goals. But in rural society, such cooperation is still confined mainly to "insiders."[42] The corresponding moral value that underpins the cooperation is mainly the value of "insiders" in the familial culture. In other words, the corresponding moral values that sustain cooperation among "outsiders," such as honesty or a social contract, are relatively undeveloped. Whether in economic or social affairs, the villagers see all "outsiders" as selfish, and in any transactions of "outsiders" with one another, the people involved simply try to make use of one another.

This election clearly shows that villagers had not yet formed any kind of contractual relationship to their mutual benefit. They were suspicious of one another. With such weak cooperation and such strong material temptations, many villagers could only think of their own personal economic gain. Meanwhile, the economic and social resources of Baixiu Village remained unfairly and unjustly distributed. The election of the Village Committee became an economic transaction. After twenty years of economic reform and the opening-up policy, it appears that the villagers' self-concept is still rooted in the moral values of the familial culture, specifically the concept of "outsiders" versus "insiders."

The moral values of the familial culture are also an obstacle to the development of individual autonomy. First, the top-down patriarchy and from it the relationship of "men decide, women follow" suppress individual choice. In this election, we can see this from the way the parents often made the decisions for their children, and husbands for their wives.

Second, the concept of "insiders" that is the cornerstone of familial moral values also deters individual autonomy. The moral values of the familial culture require that "insiders" be united and help one another. Once a branch, a *fang,* or a family has decided to back a certain candidate, every member of the family should follow that decision. Wang Zhenqiu is a good example. He did not like the secretary and did not want to vote for him for chairman of the Village Committee, yet the four *fang* of the Dezu had decided to support the secretary, so as a member of the Dezu he followed their decision and suppressed his own individual will.

Ah Chuang's views show even more vividly how the moral values of the familial culture affected the rational choice of an individual in his social dealings. Ah Chuang's elders told him that if someone from their own fraternal family or branch ran for office, no matter how lacking in ability, the family or branch members would select him. Ah Chuang said they were faced with two candidates—one an "insider," another an "outsider." Although the "insider" was inferior in many ways to the other person, they had to select the "insider." This is a duty expected of all familial members. It clearly demonstrates that when faced with choices in social and public affairs, the familial moral value of "insiders" overrides the concepts of fairness and justice. Individual choice is subject to the larger self of the lineage.

When a family member tries to ignore the familial moral value of "insider" and makes a rational and personal choice, he or she faces moral censure. Ah Sheng and Ah Zhao, who openly opposed the secretary, were regarded as being disrespectful to the elders. Worse, they were criticized by some Dezu members for having "selfish motives" and were called "rebels" and "running dogs." As a result of such censure, some villages did not reveal their opposition to the choice of the elders. Wang Guizhen's views were different from those of her elders. But she kept them to herself, and cast her vote according to her own will. The moral value of "insiders" in familism suppresses the development of individual choices.

After the Communists came to power in China, they carried out earthshaking political reforms to try to instill in peasants a collective civic sense.[43] However, thirty years of political reform have not changed the deep clan structure of the rural society.[44] Since the implementation of the economic reforms and the opening-up policy, the notion of individual autonomy in economic and social affairs has begun to sprout. Many people think that economic development will bring about the democratization of the rural village. From the first election of the Village Committee in Baixiu Village, however, we can see that a sense of political citizenship among the Chinese people is still very weak and the individual autonomy that is present in Western democratic societies has not yet developed. At the political level, the self-concept of the Chinese is still rooted in the moral values of the familial culture.

Notes

Preface

1. The following books use a "thick description" method to describe the development of Chinese villages in the post-1949 era: Huang Shumin, *The Spiral Road: Change in a Chinese Village Through the Eyes of a Communist Party Leader* (London: Westview, 1989); Anita Chan, Richard Madsen, and Jonathan Unger, *Chen Village under Mao and Deng*, 2d ed. (Berkeley: University of California Press, 1992); and Richard Madsen, *Morality and Power in a Chinese Village* (Berkeley: University of California Press, 1984). These research studies report in detail the changes in contemporary Chinese villages.

2. Names of the villages, people, and some of the places are fictitious.

Chapter 1

1. For discussion of changes in behavior and family structure of the overseas Chinese and Mainlanders, see Qiao Jian, ed., *Zhongguo Jiating Ji Qi Bianqian* (Chinese Families and Their Changes) (Hong Kong: Chinese University of Hong Kong, Hong Kong Institute of Asia-Pacific Studies, Research Monograph RM4, 1991); and Hendrick Serrie, "The Familial and the Familiar: Constancy and Variation in Chinese Culture with Reference to HSC Attributes," *Journal of Contemporary Family Studies* 16 (1985): 271–292.

2. For discussion of these two concepts, see Fei Xiaotong, *Xiangtu Zhongguo* (Rural China) (Shanghai: Guancha She, 1947), 22–33; Francis L.K. Hsu, "Chinese Kinship and Chinese Behaviour," in *China in Crisis*, vol. 1, bk. 2, *Chinese Heritage and the Commercial Political System*, ed. Ho Ping-ti and Tsou Tang (Chicago: University of Chicago Press, 1968). For further explanation of these two concepts and their applications in social research, see Li Yih-yuan, *Wenhua De Tuxiang* (The Images of Culture), 2 vols. (Taipei: Yunchen Wenhua Shiye Gongsi, 1992), chaps. 2 and 4; Li Yih-yuan, "Zhongguoren De Jiating Yu Jia De Wenhua" (The Family and the Familial Culture of the Chinese People), in *Zhongguoren: Guannian Yu Xingwei* (The

Chinese People: Viewpoints and Behavior), ed. Wen Chongyi and Xiao Xinhuang (Taipei: Juliu Tushu Gongsi, 1988).

3. Hsu, "Chinese Kinship and Chinese Behaviour," 584–587.

4. See Li Yih-yuan, "Liu Hemu De 'Fuzi' Fuqi" (The "Father and Son" Kind of Husband and Wife of Liu Hemu), and "Seqing Wenhua De Genyuan" (The Roots of the Sex Culture) in *Wenhua De Tuxiang* (The Images of Culture), vol. 1.

5. Francis L.K. Hsu has written profound and detailed discussions on this issue. See Francis L.K. Hsu, *Americans and Chinese* (Garden City, N.Y.: Doubleday Natural History Press, 1972), pt. I.

6. For further discussion of this question, see Sulamith H. Potter and Jack M. Potter, "The Cultural Construction of Emotion in Rural Chinese Social Life," chap. 9 in *China's Peasants: The Anthropology of a Revolution* (Cambridge: Cambridge University Press, 1990), 180–195.

7. Many Western scholars who study contemporary Chinese women hold a similar viewpoint. For more representative studies in this respect, see Harriet Evans, *Women and Sexuality in China* (Cambridge: Polity Press, 1997); Christina K. Gilmartin, Gail Hershatter, Lisa Rofel, and Tyrene White, eds., *Engendering China* (Cambridge, Mass.: Harvard University Press, 1994); Margery Wolf, *Revolution Postponed: Women in Contemporary China* (Stanford, Calif.: Stanford University Press, 1985); Judith Stacey, *Patriarchy and Socialist Revolution in China* (Berkeley: University of California Press, 1983); Li Xiaojiang, Zhu Hong, Dong Xiuyu, eds., *Xingbie Yu Zhongguo* (Gender and China) (Beijing: Xinzhi Sanlian Shudian, 1994).

8. See the works of Evans, Wolf, and Stacey. Woon Yuen-fong has described and analyzed this issue in detail in "From Mao to Deng: Life Satisfaction among Rural Women in an Emigrant Community in South China," *Australian Journal of Chinese Affairs* 25 (1990): 139–169.

9. To understand more about the history and significant viewpoints of the Western women's liberation movements, see Olive Banks, *Faces of Feminism* (Oxford: Basil Blackwell, 1986).

10. For more in-depth and systematic discussions on these questions, see Susan J. Hekman, *Gender and Knowledge: Elements of a Postmodern Feminism* (Cambridge: Polity Press, 1990).

11. For more concise discussion of these issues, see Seyla Benhabib and Drucilla Cornell, eds., "Introduction," *Feminism as Critique* (Cambridge: Polity Press, 1987); Shi Zhiyu and Quan Xiang, *Nuxing Zhuyi De Zhengzhi Pipan: Shui De Zhishi? Shui De Guojia?* (Political Criticism of Feminism: Whose Knowledge? Whose Country?) (Taipei: Zheng Zhong, 1994), chaps. 1 and 2.

12. See Seyla Benhabib, Judith Butler, Drucilla Cornell, and Nancy Fraser, *Feminist Contentions: A Philosophical Exchange* (London: Routledge & Kegan Paul, 1995); also see note 11 above.

13. None of the following recent studies of female social status discuss this particular issue: Sha Jicai, ed., *Dangdai Zhongguo Funu Diwei* (The Social Status of Contemporary Chinese Women) (Beijing: Beijing University Press, 1995); Xiong Yu, *Mian Dui 21 Shiji De Xuanze* (Facing the Choices of the Twenty-first Century) (Tianjin: Tianjin Renmin Press, 1993); and Chinese Women's Social Status Research Team, *Zhongguo Funu Shehui Diwei Gaiguan* (An Overview of Women's Social Status in China) (Beijing: Zhongguo Funu Press. 1993).

14. For discussions of these questions from various perspectives as well as different facets of reality, see Tu Wei-ming, ed., *China in Transformation* (Cambridge,

Mass.: Harvard University Press, 1993); Wu Guoguang, ed., *Guojia, Shichang Yu Shehui* (State, Market, and Society) (Hong Kong: Oxford University Press, 1994).

15. For a detailed in-depth discussion on this question, see F.A. Hayek, *The Constitution of Liberty* (London: Routledge & Kegan Paul, 1960). For an explanation that uses the concept of "autonomy" in connection with traditional Chinese thought, see Donald J. Munro, "Introduction," in *Individualism and Holism: Studies in Confucian and Taoist Values*, ed. Donald J. Munro (Ann Arbor: University of Michigan Press, 1985), 11–14.

16. For a more in-depth and insightful discussion that compares this question with Western values in the context of a modern society, see Yu Yingshi, "Cong Jiazhi Xitong Kan Zhongguo Wenhua De Xiandai Yiyi" (The Modern Meaning of Chinese Culture from the Point of View of Value System) in *Zhongguo Sixiang Chuantong De Xiandai Quanshi* (A Modern Interpretation of Traditional Chinese Philosophy) (Taipei: Lianjing Chuban Shiye Gongsi, 1987); and Tu Wei-ming, *Renxing Yu Ziwo Xiuyang* (Human Nature and Moral Self-Cultivation) (Taipei: Lianjing Chuban Shiye Gongsi, 1992).

17. Ibid.

18. Ibid.

19. This viewpoint is generally held by contemporary and modern anti-traditionalists. For more in-depth and detailed discussion, see Liu Zaifu and Lin Gang, *Chuantong Yu Zhongguoren* (Tradition and the Chinese) (Hong Kong: Sanlian Bookstore [Hong Kong] Company Limited, 1988).

20. Hsu, "Chinese Kinship and Chinese Behaviour."

21. See Li Yih-yuan, "Liu Hemu De 'Fuzi' Fuqi" (The 'Father and Son' Kind of Husband and Wife of Liu Hemu), and "Seqing Wenhua De Genyuan" (The Roots of the Sex Culture).

22. Beginning with John Locke, Adam Smith, and others, the philosophical, political, and economic traditions of British imperialism always held that the development of individual freedom and autonomy is closely related. There is abundant literature on this subject written by Western scholars, among which Hayek has published the most representative studies: *The Constitution of Liberty*, and *Law, Legislation and Liberty*, 3 vols. (London: Routledge & Kegan Paul, 1973–79). The arguments in the short analysis on this question in this chapter are arguments commonly known in the Western academic world, and the analysis is mainly based on Hayek's theory.

23. For discussion, see Geoffrey Hawthorn, *Enlightenment and Despair: A History of Social Theory*, 2d ed. (Cambridge: Cambridge University Press, 1987); Steven Seidman, *Liberalism and the Origins of European Social Theory* (Berkeley: University of California Press, 1983); and Anthony Giddens, *Capitalism and Modern Social Theory: An Analysis of the Writings of Marx, Durkheim and Max Weber* (Cambridge: Cambridge University Press, 1971).

24. See Seyla Benhabib, *Critique, Norm and Utopia: A Study of the Foundations of Critical Theory* (New York: Columbia University Press, 1986); Paul Connerton, *The Tragedy of Enlightenment: An Essay on the Frankfurt School* (Cambridge: Cambridge University Press, 1980); and Martin Jay, *The Dialectical Imagination: A History of the Frankfurt School and Institute of Social Research, 1923–1950* (Toronto: Little, Brown, 1973).

25. For further discussion of these questions, see Elizabeth Anderson, *Value in Ethics and Economics* (Cambridge, Mass.: Harvard University Press, 1993); Charles Taylor, *The Ethics of Authenticity* (Cambridge, Mass.: Harvard University Press, 1991);

Robert Bellah et al., *The Good Society* (New York: Knopf, 1991); Jurgen Habermas, *The Theory of Communicative Action*, vol. 2, trans. Thomas McCarthy (Boston: Beacon Press, 1987); and Yuen Sun-pong, *Pipan Lixing, Shehui Shijian, Yu Xianggang Kunjing* (Critical Reason, Social Practice, and the Dilemma of Hong Kong) (New Jersey: Global Publishing Co., 1997).

26. See Yuen Sun-pong, "Pipan Quanshilun De Lilun Jichu" (The Theoretical Basis of Critical Hermeneutics), in *Pipan Quanshilun Yu Shehui Yanjiu* (Critical Hermeneutics and Social Research), ed. Yuen Sun-pong (River Edge, N.J.: Global Publishing Co., 1993); and Yuen Sun-pong, *Pipan Lixing, Shehui Shijian Yu Xianggang Kunjing* (Critical Reason, Social Practice, and the Dilemma of Hong Kong), chaps. 1 and 2.

27. For discussions, see Terry Johnson, Christopher Dandeker, and Clive Ashworth, *The Structure of Social Theory* (London: Macmillan, 1984), chap. 3; Susan J. Hekman, *Max Weber and Contemporary Social Theory* (Notre Dame: University of Notre Dame Press, 1983); Margaret M. Poloma, *Contemporary Sociological Theory* (London: Macmillan, 1979), pt. II; and Richard J. Bernstein, *The Restructuring of Social and Political Theory* (Oxford: Basil Blackwell, 1976), pt. III.

28. The three authors of this book live in the relatively Westernized city of Hong Kong.

29. See Yuen Sun-pong, "Pipan Quanshilun De Lilun Jichu" (The Theoretical Basis of Critical Hermeneutics).

Chapter 2

1. In fact, we have continued our research on Baixiu Village since 1997. But the fieldwork has been less frequent. The data collected after 1997 are used for comparison with work done in the northern villages.

Chapter 3

1. See William L. Parish and Martin K. Whyte, *Village and Family in Contemporary China* (Chicago: University of Chicago Press, 1978), chap. 10.

2. See Sulamith H. Potter and Jack M. Potter, *China's Peasants: The Anthropology of a Revolution* (Cambridge: Cambridge University Press, 1990), 199–200.

3. Ibid., 256–257.

Chapter 7

1. Judith Stacey uses secondary data to study the patriarchal structure of Chinese society; see her book, *Patriarchy and Socialist Revolution in China* (Berkeley: University of California Press, 1983). There are other feminists who employ the concept of patriarchy to conduct research on peasant women in China. See Margery Wolf, *Revolution Postponed: Women in Contemporary China* (Stanford, Calif.: Stanford University Press, 1985); Kay Ann Johnson, *Women, the Family and Peasant Revolution in China* (Chicago: University of Chicago Press, 1983); and Phyllis Andors, *The Unfinished Liberation of Chinese Women, 1949–1980* (Bloomington: Indiana University Press, 1983).

2. Although the research perspective in this period is generally biased, it lays the foundation for later studies on peasant women in China. See the publications in note 1.

3. For Western women's liberation movements, see Olive Banks, *Faces of Feminism* (Oxford: Basil Blackwell, 1986).

4. See Joanna de Groot and Mary Maynard, eds., *Women's Studies in the 1990s: Doing Things Differently?* (New York: St. Martin's Press, 1993).

5. See Fatmagul Berktay, "Looking from the 'Other' Side: Is Cultural Relativism a Way Out?" in de Groot and Maynard, eds., *Women's Studies in the 1990s: Doing Things Differently?*

6. See, for example, Zhu Ling, *Rural Reform and Peasant Income in China: The Impact of China's Post-Mao Rural Reforms in Selected Areas* (Basingstoke, Hampshire, U.K.: Macmillan, 1991); Cheng Yuk Shing, *Modelling Peasant Household Behaviour Under Double-track Pricing System in China* (Hong Kong: Business Research Centre, Hong Kong Baptist College, 1992); Daniel Kelliher, *Peasant Power in China: The Era of Rural Reform, 1979–1989* (New Haven: Yale University Press, 1992). In addition, there have been studies on village women in the 1990s, such as Ellen Judd, *Gender and Power in Rural China* (Stanford, Calif.: Stanford University Press, 1994). Most of these studies are based on the state-society relationship. This research belongs to the field of political science or political sociology. We acknowledge the importance of this research, but we argue that without an in-depth cultural study as the foundation, it is difficult to find a valid analytical basis in the state-society relationship. Because of space limitations, we will not discuss this viewpoint here in further detail. Our purpose is only to emphasize the importance of studying Chinese rural women from the cultural perspective.

7. See the publications in note 6 and Ellen Judd's *Gender and Power in Rural China* as an example.

8. See Charles Taylor, *Sources of the Self: The Making of the Modern Identity* (Cambridge, Mass.: Harvard University Press, 1989).

Chapter 8

1. "Zhonggong Zhongyang Guanyu Jiakuai Nongye Ruogan Wenti De Jueding" (Decisions of the Central Committee of the Chinese Communist Party on Certain Questions Concerning How to Speed Up the Modernization of Agriculture) (September 28, 1979, Zhongguo Gongchandang De Shiyi Jie Zhongyang Weiyuanhui di Si Ci Quanti Huiyi Tongguo [Passed at the Fourth General Meeting of the Eleventh Central Committee of the Chinese Communist Party]), in *Xin Shiqi Nongye He Nongcun Gongzuo Zhongyao Wenxian Xuanbian* (Selected Important Documents on Agriculture and Rural Works in the New Era), ed. Zhonggong Zhongyang Wenxian Yanjiu Zhongxin (Chinese Communist Party Literature Research Center) (Beijing: Zhongyang Wenxian Chuban She, 1992).

2. Pan Zuodi, Li Derong, and Zhong Runsheng, "Xiangzhen Qiye Yao Zai Gaige Kaifang Zhong Qiu Fazhan" (The Village–Town Enterprises Must Seek Development in Reform and Opening Up), in *Guangdong Nongcun Gaige Yanjiu* (Study of the Reform of Guangdong Villages), ed. Pan Zuodi et al. (Guangzhou: Huanan Ligong Daxue Press, 1992), 396–397.

3. For development of agricultural education and education in Chinese villages, see Zhonghua Renmin Gonghe Guo Nongye Bu (Ministry of Agriculture, People's Republic of China), *Jianshe Zhongguo Tese De Xiandaihua Nongye* (Building a Modernized Agriculture with Chinese Characteristics) (Beijing: Nongye Chuban She, 1992), 419–460; and Dwight Perkins and Shahid Yusuf, *Rural Development in China* (London: Johns Hopkins University Press, 1984), 161–193.

4. For details, see Elisabeth Croll, *From Heaven to Earth: Images and Experi-

ences of Development in China (London: Routledge, 1994), 187–197.

5. The materials in this paragraph are extracted mainly from Wang Yongping, ed., *Xin Zhongguo Dashi Dian* (Almanac of New China) (Beijing: Zhongguo Guoji Guangbo Chuban She, 1992), 365–367; and Margery Wolf, *Revolution Postponed: Women in Contemporary China* (Stanford, Calif.: Stanford University Press, 1985), chap. 10. Supplementary reference materials are found in Croll, *From Heaven to Earth*, chaps. 8 and 9.

6. State Statistical Bureau, PRC, comp., *Statistical Yearbook of China, 1985* (English ed.) (Hong Kong: Economic Information Agency and Beijing: China Statistical Information and Consultancy Service Center, 1985), 185.

7. Croll describes six phases to conclude the evolution of the policies within the decade from 1979 to 1989. See Croll, *From Heaven to Earth*, 188–192.

8. Some of the materials used in this paragraph are taken from Yao Ruobing, ed., *Zhongguo Jiaoyu* (Education in China) *1949–1982* (Hong Kong: Huafeng Shuju, 1984), 6–7; and Wang Yongping, ed., *Xin Zhongguo Dashi Dian* (Almanac of New China), 480–482. See also the following supplementary reference materials: Cheng Kai Ming, *Zhongguo Jiaoyu Gaige: Jinzhan, Juxian, Qushi* (Education Reform in China: Progress, Limitations, Trends) (Hong Kong: Commercial Press, 1992); and John N. Hawkins, *Education and Social Change in the People's Republic of China* (New York: Praeger, 1983).

Chapter 9

1. See Sulamith H. Potter and Jack M. Potter, *China's Peasants: The Anthropology of a Revolution* (Cambridge: Cambridge University Press, 1990), chap. 9.

Chapter 10

1. See Harriet Evans, *Women and Sexuality in China: Dominant Discourses of Female Sexuality and Gender Since 1949* (Cambridge: Polity Press, 1997); Katherine Carlitz, "Desire, Danger, and the Body: Stories of Women's Virtue in Late Ming China," in *Engendering China: Women, Culture and the State*, ed. Christina K. Gilmartin, Gail Hershatter, Lisa Rofel, and Tyrene White (Cambridge, Mass.: Harvard University Press, 1994).

2. Evans, *Women and Sexuality in China: Dominant Discourses of Female Sexuality and Gender Since 1949*, chap. 1; and Zheng Sili, *Zhongguo Xing Wenhua: Yige Qiannian Bujie Zhi Jie* (Sex Culture in China: A Knot Left Untied for a Thousand Years) (Beijing: Zhongguo Duiwai Fanyi Chuban Gongsi, 1994).

Chapter 11

1. See Li Yih-yuan, "Seqing Wenhua De Genyuan" (The Roots of the Sex Culture), in *Wenhua De Tuxiang* (The Images of Culture), vol. 1 (Taipei: Yun Chen, 1992).

2. For discussions of traditional Chinese gender relationships and standards of sexual behavior, see Li Meizhi, "Liang Xing Guanxi De Shehui Shengwuxue Yuanxing Zai Chuantong Zhongguo Yu Jinri Taiwan De Biaoxian" (The Expression of the Sociobiological Model of Gender Relationships in Traditional China and Contemporary Taiwan) and Yang Zhongfang, "'Shehui/Wenhua/ Lishi' De Kuangjia Zai Nali" (Where Is the Framework for Society/Culture/History?) in *Bentu Xinlixue Yanjiu* (Indigenized

Psychological Studies), vol. 5 (1996); and Zheng Sili, *Zhongguo Xing Wenhua: Yige Qiannian Bujie Zhi Jie* (Sex Culture in China: A Knot Left Untied for a Thousand Years) (Beijing: Zhongguo Duiwai Fanyi Chuban Gongsi, 1994).

3. For discussions of these questions in connection with the issue of face (*mianzi*), see Zhai Xuewei, *Zhongguoren De Lianmian Guan* (The Chinese People's Idea of Faces) (Taipei: Guiguan, 1994); Ambrose King, "'Mian,' 'Chi' Yu Zhongguoren Xingwei Zhi Fenxi" ("Face," "Shame," and an Analysis of Chinese People's Behavior), in *Zhongguo Shehui Yu Wenhua* (Chinese Society and Culture) (Hong Kong: Oxford University Press, 1992); and Huang Guangguo, "Renqing Yu Mianzi: Zhongguoren De Quanli Youxi" (Human Feelings and Face: The Chinese People's Power Game), in *Zhongguoren De Quanli Youxi* (The Chinese People's Power Game), ed. Huang Guangguo (Taipei: Juliu, 1988).

Chapter 13

1. Gordon Chu and Yanan Ju, *The Great Wall in Ruins* (New York: State University of New York Press, 1993).

2. See Li Zehou and Liu Zaifu, *Gaobie Geming* (Farewell to Revolution) (Hong Kong: Cosmos Book Co., 1995).

3. Wu Guoguang, "Biange Zhong De Guojia Nengli: Yige Beilun He San Zhong Nengli" (The State Ability in Reform: A Paradox and Three Types of Abilities), in *Guojia, Shichang Yu Shehui* (State, Market, and Society), ed. Wu Guoguang (Hong Kong: Oxford University Press, 1994).

4. Huang Zongzhi, *Changjiang Sanjiaozhou Xiaonong Jiating Yu Xiangcun Fazhan: 1350–1988* (The Development of Peasant Families and Villages in the Yangtze River Delta) (Hong Kong: Oxford University Press, 1994); Huang Shumin, foreword to *The Spiral Road: Change in a Chinese Village Through the Eyes of a Communist Party Leader* (London: Westview, 1989).

5. Li Zehou and Liu Zaifu, *Gaobie Geming* (Farewell to Revolution), chap. 2.

6. See Elizabeth Anderson, *Value in Ethics and Economics* (Cambridge, Mass.: Harvard University Press, 1993); Yuen Sun-pong, *Pipan Lixing, Shehui Shijian Yu Xianggang Kunjing* (Critical Reason, Social Practice, and the Dilemma of Hong Kong) (River Edge, N.J.: Global Publishing Co., 1997), chaps. 3 and 4.

7. Yu Yingshi, *Zhongguo Wenhua Yu Xiandai Bianqian* (Chinese Culture and Its Modern Transformation) (Taipei: San Min, 1992), chaps. 12 and 13; Yuen Sun-pong, *Pipan Lixing, Shehui Shijian Yu Xianggang Kunjing* (Critical Reason, Social Practice, and the Dilemma of Hong Kong), chap. 6.

8. Charles Taylor, *The Ethics of Authenticity* (Cambridge, Mass.: Harvard University Press, 1991), chap. 3.

9. The point is developed from Charles Taylor and Sulamith Potter and Jack Potter. See Charles Taylor, "What's Wrong with Negative Liberty," chap. 8 in *Philosophy and the Human Sciences: Philosophical Papers*, vol. 2 (Cambridge: Cambridge University Press, 1985), 211–229; Sulamith H. Potter and Jack M. Potter, "The Cultural Construction of Emotion in Rural Chinese Social Life," chap. 9 in *China's Peasants: The Anthropology of a Revolution* (Cambridge: Cambridge University Press, 1990), 180–195.

10. Emile Durkheim, *The Division of Labour in Society*, trans. W.D. Halls (London: Macmillan, 1984).

11. See Qiao Jian, ed., *Zhongguo Jiating Ji Qi Bianqian* (Chinese Families and

Their Changes) (Hong Kong: Chinese University of Hong Kong, Hong Kong Institute of Asia-Pacific Studies, Research Monograph RM4, 1991).

12. Gordon Chu and Yanan Ju, *The Great Wall in Ruins*, chap. 12.

Chapter 14

1. An outstanding example in this regard is E.R. Babbie, *The Practice of Social Research* (Belmont, Calif.: Wadsworth, 1994).

2. Anthony Giddens also accepted this view. See Anthony Giddens, *Sociology: A Brief But Critical Introduction* (London: Macmillan, 1982), 1–2.

3. See Mary Hesse, "Introduction," in *Revolutions and Reconstructions in the Philosophy of Science* (Sussex: Harvester Press, 1980). But the discussion here differs slightly from Hesse's analysis.

4. For a detailed discussion on the basic phenomena, see Charles Taylor, "Interpretation and Science of Man," *Philosophy and the Human Sciences: Philosophical Papers*, vol. 2 (Cambridge: Cambridge University Press, 1985).

5. See James Bohman, *New Philosophy of Science: Problems of Indeterminacy* (Cambridge: Polity Press, 1990); Richard J. Bernstein, *Beyond Objectivism and Relativism* (Oxford: Basil Blackwell, 1983); Hesse, *Revolutions and Reconstructions in the Philosophy of Science.*

6. See Mary Hesse, "Theory and Value in the Social Sciences," in *Action and Interpretation*, ed. Christopher Hookway and Philip Petti (Cambridge: Cambridge University Press, 1978).

7. See note 5.

8. The discussion here is based on Jurgen Habermas's consensus theory of truth. See Jurgen Habermas, "A Postscript to Knowledge and Human Interests," in *Knowledge and Human Interests*, trans. J. Shapiro (Boston: Beacon Press, 1971). For more detailed discussion, see Yuen Sun-pong, *Pipan Lixing, Shehui Shijian Yu Xianggang Kunjing* (Critical Reason, Social Practice, and the Dilemma of Hong Kong) (River Edge, N.J.: Global Publishing Co., 1997), chap. 2.

9. James S. Coleman argues that we would be able to obtain a more thorough explanation about the social phenomenon if we viewed it from the perspective of individual interests. See James S. Coleman, *Individual Interests and Collective Action* (Cambridge: Cambridge University Press, 1986), 1. Coleman tries to present the method of thick description used in social research with reference to the realization or lack of realization of human interests. But he does not provide an adequate answer to the question of why the reference to human interests or purposes would render a social phenomenon intelligible and why people would ask no more questions about it. Nor does he go further into the method of thick description. It seems that it is mainly due to the restrictions of the metatheoretical or epistemological assumptions of positivism that Coleman is unable to go further into the issue of rich or deep understanding. Indeed, for Coleman, all social explanations are ultimately reduced to a causal explanation.

10. For discussion on how to apply these viewpoints to social research, see Phil Francis Carspecken, *Critical Ethnography in Educational Research: A Theoretical and Practical Guide* (London: Routledge, 1996); Raymond A. Morrow, *Critical Theory and Methodology* (London: Sage, 1994); and Elliot G. Misher, *Research Interviewing: Context and Narrative* (Cambridge, Mass.: Harvard University Press, 1986).

11. For more in-depth and stimulating discussion on this question, see Taylor,

"Interpretation and Science of Man"; and John Searle, *The Construction of Social Reality* (New York: Free Press, 1995), chaps. 1 and 3.

12. For further discussion see Yuen Sun-pong, "Pipan Quanshilun De Lilun Jichu" (The Theoretical Basis of Critical Hermeneutics), in *Pipan Yu Quanshilun Shehui Yanjiu* (Critical Hermeneutics and Social Research) (River Edge, N.J.: Global Publishing Co. Inc., 1993); and Bohman, *New Philosophy of Science: Problems of Indeterminacy*.

13. The principle of "value relevance" was initially suggested by Heinrich Rickert. Later, Max Weber applied this principle to his interpretive sociology. See Heinrich Rickert, *The Limits of Concept Formation in Natural Science: A Logical Introduction to the Historical Sciences*, ed. and trans. Guy Oakes (Cambridge: Cambridge University Press, 1986), 88–89.

14. It should be mentioned that Weber's introduction of values into social research is not as simple as stated before. Many scholars point out that Weber's opinion on this issue differs greatly from positivism. See Christopher G.A. Bryant, *Positivism in Social Theory and Research* (London: Macmillan, 1985), chap. 3. However, in light of the main purpose of this book, we are borrowing Weber's discussion of "value relevance" to initiate a discussion about value involvement in the positivist conception of knowledge.

15. For discussion of this question, see Max Weber, "'Objectivism' in Social Science and Social Policy," in *The Methodology of the Social Sciences*, trans. and ed. Edward A. Shils and Henry A. Finch (New York: Free Press, 1949).

16. See Yuen Sun-pong, "Pipan Quanshilun De Lilun Jichu" (The Theoretical Basis of Critical Hermeneutics), 19–28.

17. We stress that the term "reasonable premise" is used to understand the reasons given by an individual that support his or her own behavior, and that such reasons are based on the related cultural network and value system. We do not necessarily agree with his or her reasons, but we accept that this is a related reason to support the individual's behavior. In fact, the premise that has been provided by the individual for his or her own behavior may not coincide with the underlying "reasonable" premise imagined by the interpreters. It should be noted that the interpreters may imagine a "reasonable" premise, and this is the condition precedent to interpreting the social phenomena.

18. This hypothesis is developed from Habermas's analysis. See Jurgen Habermas, *The Theory of Communicative Action*, vol. 1, trans. Thomas McCarthy (Boston: Beacon Press, 1984), 115–116.

19. The discussion of the "strong thesis of value involvement" here is based on parts of Yuen Sun-Pong's essay, "Shehui Gongzuo Lilun Yu Shijian" (The Theory and Practice of Social Work), which is forthcoming. See also Yuen Sun-pong, "Pipan Quanshilun De Lilun Jichu" (The Theoretical Basis of Critical Hermeneutics), 19–28.

20. The analysis here is based on Habermas's theory. See Jurgen Habermas, *Theory and Practice*, trans. John Viertel (Boston: Beacon Press, 1974), 18; and Yuen Sun-pong, *Pipan Lixing, Shehui Shijian Yu Xianggang Kunjing* (Critical Reason, Social Practice, and the Dilemma of Hong Kong), chap. 2.

21. See Yuen Sun-pong, *Pipan Lixing, Shehui Shijian Yu Xianggang Kunjing* (Critical Reason, Social Practice, and the Dilemma of Hong Kong), chap. 2.

22. In the 1950s and 1960s, this topic led to general discussions among academics in sociology and philosophy in Britain and the United States. For the major theses of

these discussions, see John O'Neill, ed., *Modes of Individualism and Collectivism* (London: Heinemann, 1973). For more recent discussions that, to a certain extent, have surpassed the preceding one, see Jeffrey C. Alexander et al., eds., *The Micro-Macro Link* (Berkeley: University of California Press, 1987); and James S. Coleman, "Social Theory, Social Research, and a Theory of Action," *American Journal of Sociology* 91 (1986): 1309–1335.

23. Habermas and Giddens are two outstanding scholars in this area.

24. For detailed discussion, see Charles Taylor, *Sources of the Self: The Making of Modern Identity* (Cambridge, Mass.: Harvard University Press, 1989), pt. I; and Charles Taylor, "The Moral Topography of the Self," in *Hermeneutics and Psychological Theory: Interpretive Perspectives on Personality, Psychotherapy, and Psychology*, ed. Stanley B. Messer, Louis A. Sass, and Robert L. Woolfolk (New Brunswick, N.J.: Rutgers University Press, 1988).

25. For more detailed explanation and analysis of this question, see Sulamith H. Potter and Jack M. Potter, *China's Peasants: The Anthropology of a Revolution* (Cambridge: Cambridge University Press, 1990), chap. 9.

26. For further discussion of these questions and the indigenization of the social sciences, see Yuen Sun-pong, "Houshe Lilun, Gerenguan, Yu Shehui Kexue Bentuhua" (Metatheory, Individualism, and Indigenization of the Social Sciences), *Zhongguo Shehui Kexue Jikan* (China Social Sciences Quarterly) 15 (1996): 163–170.

Chapter 15

1. Francis L.K. Hsu, *Rugged Individualism Reconsidered: Essays in Psychological Anthropology* (Knoxville: University of Tennessee Press, 1983), 221–224.

2. See Elisabeth Croll, *From Heaven to Earth: Images of Development in China* (London and New York: Routledge, 1994), 199.

Chapter 16

1. *Cunmin Weiyuanhui Zuzhifa Jianghua* (Introduction to the PRC Village Committee Organic Law) (Beijing: Legal Publishing House of China, 1999), 1 and 21.

2. *Zhonghua Renmin Gongheguo Cunmin Weiyuanhui Zuzhifa* (PRC Village Committee Organic Law), 1998, Article IV; Lianjiang Li and Kevin J. O'Brien, "The Struggle over Village Elections," in *The Paradox of China's Post-Mao Reforms*, ed. Merle Goldman and Roderick MacFarquhar (Cambridge, Mass.: Harvard University Press, 1999), 129.

3. Amy Epstein, "Village Elections in China: Experimenting with Democracy," in *China's Economic Future: Challenges to U.S. Policy*, ed. Joint Committee, Congress of the United States (Armonk, N.Y.: M.E. Sharpe, 1997), 420; Zhenyao Wang, "Village Committees: The Basis for China's Democracy," in *Cooperative and Collective in China's Rural Development: Between State and Private Interests*, ed. Eduard B. Vermeer, Frank N. Pieke, and Woei Lien Chong (London: M.E. Sharpe, 1998), 254–255; Emerson M.S. Niou, "Village Elections: Roots of Democratization in China," paper presented at the Conference on Village Self-Governance in China, December 1–2, 2000.

4. Kevin O'Brien, "Villagers, Elections, and Citizenship in Contemporary China," *Modern China* 27 (2001): 423–426.

5. Robert A. Pastor and Qingshan Tan, "The Meaning of China's Village Elections," *China Quarterly* 162 (2000): 490–512.

6. Tianjian Shi, "Cultural Values and Democracy in the People's Republic of China," *China Quarterly* 162 (2000): 540–541; Lucian W. Pye, *The Spirit of Chinese Politics* (Cambridge, Mass.: Harvard University Press, 1992), 93.

7. Wang Zhongtian and Zhan Chengfu, eds., *Xiangcun Zhengzhi: Zhongguo Cunmin Zizhi De Diaocha Sikao* (Village Politics: Reflections on Surveys on Chinese Villagers' Self-Government) (Nanchang: Jiangxi People's Press, 1999), 10; Jean Oi, "Economic Development, Stability and Democratic Village Self-Governance," in *China Review 1996*, ed. Maurice Brosseau, Suzanne Pepper, and Tsang Shu-ki (Hong Kong: Chinese University Press, 1996), 135; Sylvia Chan, "Research Notes on Village Committee Elections: Chinese-style Democracy," *Journal of Contemporary China* 7 (1998): 2–3; Kevin J. O'Brien, "Implementing Political Reform in China's Villages," *Australian Journal of Chinese Affairs* 32 (1994): 41–42.

8. Wang Zhongtian and Zhan Chengfu, eds., *Xiangcun Zhengzhi: Zhongguo Cunmin Zizhi De Diaocha Sikao* (Village Politics: Reflections on Surveys on Chinese Villagers' Self-Government), 3–4.

9. Pou Xing-zu et al., *Zhonghua Renmin Gongheguo Zhengzhi Zhidu* (The Political System of the People's Republic of China) (Hong Kong: Joint Publishing Co., 1995), 391–392; Y.S. Cheng and H.F. Tse, eds., *Dangdai Zhongguo Zhengfu* (Contemporary Chinese Government) (Hong Kong: Cosmos Books, 1992), 343; Wang Zhongtian and Zhan Chengfu, eds., *Xiangcun Zhengzhi: Zhongguo Cunmin Zizhi De Diaocha Sikao* (Village Politics: Reflections on Surveys on Chinese Villagers' Self-Government), 43.

10. Wei Qingquan et al., *Shijizhijiaode Zhujiang Sanjiaozhou Xingzheng Quhua* (Administrative Planning in the Pearl River Delta at the Turn of the Century) (Guangzhou: Guangdong Cartographic Publishing House, 1997), 42–46.

11. David Zweig, *Agrarian Radicalism in China, 1968–1981* (Cambridge, Mass.: Harvard University Press, 1989), 181.

12. Jean C. Oi, "Two Decades of Rural Reform in China: An Overview and Assessment," *China Quarterly* 159 (1999): 626; Elizabeth J. Perry, "Rural Collective Violence: The Fruits of Recent Reforms," in *The Political Economy of Reform in Post-Mao China*, ed. Elizabeth J. Perry and Christine Wong (Cambridge, Mass.: Harvard University Press, 1985), 179.

13. Daniel Kelliher, "The Chinese Debate over Village Self-Government," *China Journal* 37 (1997): 71.

14. Daniel Kelliher's study points out that after 1984 such problems became more serious daily. See Kelliher, "The Chinese Debate over Village Self-Government," 72; Richard J. Latham, "The Implications of Rural Reforms for Grass-Roots Cadres," in *The Political Economy of Reform in Post-Mao China*, 170–172; Liangjiang Li and Kevin J. O'Brien, "Villagers and Popular Resistance in Contemporary China," *Modern China* 22 (1996): 1; Epstein, "Village Elections in China: Experimenting with Democracy," 411.

15. Kuo-cheng Sung, "Peasant Unrest in Szechwan and Mainland China's Rural Problems," *Issues and Studies* 27 (1993): 129; Epstein, "Village Elections in China: Experimenting with Democracy," 416.

16. Perry, "Rural Collective Violence: The Fruits of Recent Reforms," 178–179.

17. Li and O'Brien, "The Struggle over Village Elections," 131–133, 138–139; Li and O'Brien, "Villagers and Popular Resistance in Contemporary China," 1; Kelliher, "The Debate over Village Self-Government," 71–73; Epstein, "Village Elections in China: Experimenting with Democracy," 411–413; Oi, "Economic Development and

Village Self-Governance"; Cheng Tongshun, *Dangdai Zhongguo Nongcun Zhengzhi Fazhan Yanjiu* (A Study of Political Development in Contemporary Chinese Villages) (Tianjin: Tianjin People's Press, 2000), 140.

18. Oi, "Economic Development and Village Self-Governance," 126–127; Kelliher, "The Chinese Debate over Village Self-Government," 68; Cheng Tongshun, *Dangdai Zhongguo Nongcun Zhengzhi Fazhan Yanjiu* (A Study of Political Development in Contemporary Chinese Villages), 136–137.

19. Kelliher, "The Debate over Village Self-Government," 67–78; Li and O'Brien, "The Struggle over Village Elections," 131–133.

20. Li and O'Brien, "The Struggle over Village Elections," 133.

21. These six provincial and provincial-level governments include Guangdong, Shanxi, Hainan, Guangxi, Yunnan, Shanghai, and Beijing.

22. Melanie Manion, "The Electoral Connection in the Chinese Countryside," *American Political Science Review* 90 (1996): 738; Li and O'Brien, "The Struggle over Village Elections," 136; Kelliher, "The Debate over Village Self-Government," 78–81; O'Brien, "Implementing Political Reform in China's Villages," 54–55.

23. *Zhonghua Renmin Gongheguo Cunmin Weiyuanhui Zuzhifa (Shixing)* (PRC Village Committee Organic Law [Trial]), 1987, Article III.

24. Kelliher, "The Debate over Village Self-Government," 78.

25. Li and O'Brien, "The Struggle over Village Elections," 136; Kelliher, "The Debate over Village Self-Government," 80.

26. O'Brien, "Implementing Political Reform in China's Villages," 56; Li and O'Brien, "The Struggle over Village Elections," 140; Kelliher, "The Debate over Village Self-Government," 79.

27. Li and O'Brien, "The Struggle over Village Elections," 138–140.

28. *Hai xuan* is explained in *Cunmin Weiyuanhui Zuzhifa Jianghua*, 69. It means to get a needle from the sea or to collect pearls from the sea. Sylvia Chan's study points out that the number of candidates who won through the *hai xuan* method was very high, so they were hard to control. The study also found that in those villages where *hai xuan* produced the list of candidates, the voting rate was very high. Chan, "Research Notes on Village Committee Elections: Chinese-style Democracy."

29. *Guangdongsheng Cunmin Weiyuanhui Xuanju Guichen* (Guangdong Province Village Committee Election Regulations) (Guangdong Province Work Guidance Team Office to Manage Village Basic-Level Administrative Structure, 1998), 36.

30. Chan, "Research Notes on Village Committee Elections: Chinese-style Democracy," 510.

31. *Guangdongsheng Cunmin Weiyuanhui Xuanju Guichen* (Guangdong Province Village Committee Election Regulations), 14; According to reports, some Party branch members or secretaries resigned from the post of Party secretary in order to run for the chairmanship of the Villager Committee. See Epstein, "Village Elections in China: Experimenting with Democracy," 419.

32. *Guangdongsheng Cunmin Weiyuanhui Xuanju Guichen* (Guangdong Province Village Committee Election Regulations), 51.

33. David Zweig's study points out that after the economic reform and opening-up policy, the right to land use became the flash point for conflicts between villagers and Party cadres. See David Zweig, "Struggling over Land in China: Peasant Resistance After Collectivization, 1966–1986," in *Everyday Reforms of Peasant Resistance*, ed. Forrest D. Colburn (Armonk, N.Y.: M.E. Sharpe, 1989), 162–163.

34. O'Brien, "Villagers, Elections, and Citizenship in Contemporary China," 425.

35. Pastor and Tan, "The Meaning of China's Village Elections," 491–512.

36. Thomas Metzger, "The Western Concept of the Civil Society in the Context of Chinese History," Hoover Essay No. 21, Hoover Institution on War, Revolution and Peace, Stanford University (1998), 12–17.

37. Xiaotong Fei, *From the Soil: The Foundations of Chinese Society*, trans. Gary G. Hamilton and Wang Zheng (Berkeley: University of California Press, 1992), 60–70.

38. Maurice Freedman, *Lineage Organization in Southern China* (London: Athlone, 1958), 46–50; Hugh Baker, *A Chinese Village: Sheung Shui* (Stanford, Calif.: Stanford University Press, 1968).

39. Richard Madsen, "The Public Sphere, Civil Society and Moral Community," *Modern China* 19 (1993): 183–199; Metzger, "The Western Concept of the Civil Society in the Context of Chinese History," 13. Metzger attempts to analyze the moral values of Chinese familism to show that this is one of the cultural factors that makes it difficult for Chinese society to develop into a civil society similar to Western society.

40. O'Brien, "Villagers, Elections, and Citizenship in Contemporary China," 424.

41. Jude Howell, "Prospects for Village Self-Governance in China," *Journal of Peasant Studies* 25 (1998): 86–111.

42. Elisabeth Croll, *From Heaven to Earth: Images and Experiences of Development in China* (London: Routledge, 1994), 172.

43. Richard Madsen, *Morality and Power in a Chinese Village* (Berkeley: University of California Press, 1984), 100.

44. Sulamith Potter and Jack Potter, *China's Peasants: The Anthropology of a Revolution* (Cambridge: Cambridge University Press, 1990), 255.

Index

Sun-pong Yuen is professor in the Department of Applied Social Sciences at the Hong Kong Polytechnic University. He is the founder of a Chinese, refereed journal, *Shehui Lilun Xuebao* (Journal of Social Theory) and currently its chief editor. Professor Yuen received recognition for research in the following areas: China's modernization, contemporary social theory, indigenization of social sciences, and social work theory and practice. His books include *Pipan Quanshilun Yu Shehui Yanjiu* (Critical Hermeneutics and Social Research) (1993), *Pipan Lixing, Shehui Shijian Yu Xianggang Kunjing* (Critical Reason, Social Practice, and the Dilemma of Hong Kong) (1997), and *Reason and Practice in Philosophical Hermeneutics and Confucianism* (forthcoming). He also coedited *Dangdai Zhongguo Nongcun Yanjiu (Shang): Lilun Tansuo* (Studies in Contemporary Rural China, vol. 1, Theoretical Exploration) (2000) and *Hermeneutics and Social Work* (forthcoming).

Pui-lam Law is a Ph.D. candidate in sociology at the University of New South Wales, Australia, and is currently a lecturer in the Department of Applied Social Sciences of the Hong Kong Polytechnic University. He coedited *Dangdai Zhongguo Nongcun Yanjiu (Shang): Lilun Tansuo* (Studies in Contemporary Rural China, vol. 1, Theoretical Exploration) (2000) and *Dangdai Zhongguo Nongcun Yanjiu (Xia): Shizheng Diaocha* (Studies in Contemporary Rural China, vol. 2, Empirical Investigation) (2000).

Yuk-ying Ho received her Ph.D. from the Hong Kong Polytechnic University. She is currently a postdoctoral fellow of the Department of Applied Social Sciences at The Hong Kong Polytechnic University, studying the self-concept of college students in Hong Kong. She is also an assistant editor of *Shehui Lilun Xuebao* (Journal of Social Theory).

Translator

Fong-ying Yu was an associate professor in the Department of Chinese and Bilingual Studies at the Hong Kong Polytechnic University, and is currently an independent language professional. He specializes in bilingual communication in English and Chinese, TESOL, language education, and translation.